THE
RIVER

THE
RIVER

Philippa Forrester

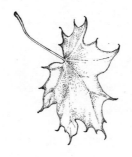

arrow books

Published by Arrow Books 2010

10 9 8 7 6 5 4 3 2 1

Copyright © Philippa Forrester 2004, 2010

Philippa Forrester has asserted her right to be identified as the author of this work
under the Copyright, Designs and Patents Act 1988

Drawings by Delia Delderfield

First published in Great Britain in 2004 by Orion
An imprint of Orion Books Ltd
Orion House, 5 Upper St Martin's Lane
London WC2H 9EA

An imprint of The Random House Group Limited

www.rbooks.co.uk

Addresss for companies within The Random House Group Limited
can be found at www.randomhouse.co.uk

The Random House Group Limited Reg. No. 954009

A CIP catalogue record for this book is available from the British Library

ISBN 978 1 84809 247 1

Mixed Sources
Product group from well-managed
forests and other controlled sources
www.fsc.org   Cert no. TT-COC-2139
© 1996 Forest Stewardship Council

The Random House Group Limited supports The Forest Stewardship Council
(FSC), the leading international forest certification organisation. All our titles that
are printed on Greenpeace approved FSC certified paper carry the FSC logo. Our
paper procurement policy can be found at www.rbooks.co.uk/environment

Printed and bound in Great Britain by CPI Bookmarque, Croydon CR0 4TD

To the light of my life Charlie,
to my darling boys Fred, Gus and Arthur
and to grumpy, lovely old Bill,
who we miss every day

CONTENTS

ACKNOWLEDGEMENTS

There are many people to whom I owe thanks, firstly to all the friends and neighbours who appear in the pages of this book and light up our lives. Any resemblance to their real selves is entirely in their own imagination. Particular thanks to Richard and Stephanie for giving us the constant run of their home and making us feel just like family.

Secondly to Julian Alexander without whom I would still be plodding along writing a diary of our lives for my own amusement. His politic and patient nature has been the perfect foil for my own less temperate one; his advice invaluable at all times.

And then I suppose I will have to thank Trevor Dolby even though he has been a complete pain for the whole time. It was his fault I had to write this book in the first place and then he kept poking his nose in and trying to tell me how to do it. As if that weren't enough, I now find he has turned himself into a friend so we have to put up with him phoning and visiting as well.

I owe Alan Samson many thanks for taking over and running with the book he was given.

Many thanks also to Delia for the charming illustrations and her patience when waiting for source material.

To Laura, our northern light, for being here long enough for me to get some work done and for putting up with our lives. You'll never know how much you mean to us. And to Hilary and Kate for keeping the wolves from the door long enough for me to get some work done.

To all the people who helped us through the making of the programme: Mike Gunton for having the bravery and creative genius to commission it! Keith Scholey for having faith in us, Jamie for putting up with week after week of endless mickey-taking and caring about it as much as we did, Nigel, Anna, Jeremy, Neill and Martyn. I hope you like what you find in the pages ahead.

To both our families, Mum, Dad and Deedaw and John, who have the wisdom to keep their mouths shut when we announce another ridiculous plan but have the love to support us through it all.

Oh . . . and the otters.

Chapter One
THE TIDE TURNS

....

Imperceptibly, inch by inch, life changes

Our life has taken a strange turn. Our dinner-table talk is full of kingfisher tales and punctuated by bird spotting, breakfast is occupied placing bets on whether a squirrel can leap the river without getting his tail wet. Home life and work life have become inseparable. Our nights are interrupted by the comings and goings of otters, our days dominated by the mood of the river, our frame of mind set by the presence or absence of wildlife. We are filming the river that flows past our house and the animals that live with us on the river bank. It feels as though we are no longer part of the normal human run of things, as though suddenly we have been transported into a world removed from anything I have known before, and some days I find I am still spinning from the shock of it. On those days I pinch myself and try to recall how this abrupt and surprising change of direction happened. How I got into such a delightful mess. Looking back I can see a few signposts which should have served to warn me, had I been paying attention.

It began when I found myself engaged to a wildlife cameraman. Tall, dark and handsome, with a penchant for whisky on cold days and a skill for delivering fantastic breakfasts, I couldn't resist him. Charlie Hamilton James walked into my life one day to do a spot of otter filming in Skye and then stayed, with his Border collie Bill.

I had met him once before, the love of my life, over a carcass. We didn't really hit it off. If I had read more Mills and Boon stories I would have realised that this is how the best relationships start – excepting the carcass, that of course is not normal. I had been sent a schedule instructing me to meet a crew at Longleat Safari Park to film wolves. It was a beautiful day; blue sky and birdsong made me smile and a long drive made me thirsty. Charlie wasn't very impressed with the minor celebrity rolling up in her sports car, an inane grin on her face, especially when she went on to steal his Ribena. So we didn't speak much that day. He thought I was arrogant and selfish and I thought he was a show-off and a bore. Charlie and I nodded our goodbyes when the work was done at the end of the day and I drove away assuming that would be the last I'd see of this rude man.

We next met when, a few months later, I was waiting at Heathrow for a crew who were late. We were meant to be catching a plane to go and film otters on the Kyle of Lochalsh which is on the west coast of Scotland next to Skye. It was to be my first visit to the area. I had often heard friends rave about how beautiful it was and so I was very excited. But the clock was ticking, we were well into check-in time, and still there was no sign of the crew. I decided to phone. I looked at the call sheet to find a number and recognised Charlie's name under 'cameraman'.

He turned up ten minutes later, a huge grin splitting his face. He had either forgotten the Ribena incident or forgiven me, and my own grumpy feelings began to evaporate as I watched him get out of the car. I had forgotten how good-looking he was, and while we checked in the many bags I kept sneaking glances. He is six foot three, with dark hair, a classic square jaw, lots of stubble and a laconic smile. But what really dazzled me were his eyes, bright blue and dancing with life and humour. I found it very difficult to stop looking into them.

We just made it to the plane on time. Once on board, after coffee and a vague attempt at a bacon baguette, we did the only thing left to do and studied the perfume catalogue. Soon we were crying with laughter at the ridiculous descriptions of the various scents on offer; there were scents which would tell the world you were in charge of your life, smells which would awaken the beast in you, aromas which would whisk you from your bathroom away to the mystical orientals for a little Eastern magic, whiffs which would turn your very pulse points into paradise for your *amour*. We congratulated the person whose job it is to string words together in such an alluring way for their grand effort but on a trip to Skye we would have been more interested in midge repellent.

After the usual shenanigans trying to hire a car which had room for us and the kit we were off out of the city and into the wonderful Scottish Highlands. I didn't mind a minute of the long drive to Skye; the scenery was so expansive and every bit as beautiful as I had been led to believe. We climbed past snow-topped peaks and a thousand different shades of green; we went through valleys where the icy mountain water tumbled down from great heights beside us and rolled by loch upon loch. The light and the weather were constantly changing as we bounced along in the car and the surface of the lochs went from grey to green to choppy to smooth. We had the stereo on loud. We were on an adventure.

Our mission? To film the otters in Kyle harbour, who have learned that there are always pickings to be had when the fishing boats return from a trip. Although normally otters are extremely shy and reluctant to come near humans, these otters are constantly on the boats, hoovering up the leftovers.

Sadly, we had to abandon the first night's filming because of stormy seas and a lack of fishing boats. Desperate to catch sight of an otter, we tried to go out in the weather, but could barely stand in the wind and rain and were soaked to the skin in seconds. But the second night was magical, fine, clear but deathly cold. Bundled in several layers of thermals and some especially warm socks called Mountain Toasties which, knowing how cold it gets in a Scottish harbour in November at night, Charlie had thoughtfully bought for all of us, I looked like the Michelin Man. Not an image I'd normally have chosen for seduction.

We arrived at the boats a little before dark so that we could settle in and be ready for the otters. The street lights along the harbour wall had just come on and the harbour itself was quiet, apart from the distant wash of waves breaking and the boats moving with the sea and bumping against the wall. It was a little windy but nothing like the night before. As quietly and as quickly as possible, we risked life and limb on the steep and slippery harbour steps carrying boxes of heavy equipment onto our boat. We rigged up lights and monitors, the director hummed and hahed over the shot and changed it a few times before he was happy, then we mumbled about the script and the kind of thing it would be best to say when the great moment arrived and the otters turned up. I wanted something a bit more meaningful than 'Wow!' but then we all agreed that 'Wow!' was probably appropriate. At that point we realised we were starving, and all that cold sea air demanded only one thing: fish and chips!

Richard the director offered to go and we remained with

the sound man Simon, who seemed more interested in the hospitality in the hotel than the film we were trying to make. He was English but now lived in Scotland because he was suspicious that England was becoming full of the 'wrong types'. Having moved to Scotland to get away from it all it seemed he could find little to be happy about there either.

'So what time will these otters come then?'

'Don't know,' Charlie replied. 'When they're ready.'

'Well, will it be before eleven?'

'If it suits them. Why?'

'The pub shuts then.'

Charlie and I exchanged looks. We knew there was a possibility that we'd be waiting all night and we had to be prepared to do just that if we wanted a good chance of getting the shots. We also knew there was a strong likelihood that they wouldn't show up at all. Otters aren't very familiar with pub closing times.

The fish and chips arrived. Richard reluctantly volunteered to keep a lookout from the dockside and retired there with his warm, soggy newspaper parcel. There is nothing like fish and chips by the sea – the smell of vinegar, salt on numb fingers and the crispy batter.

I looked up to see how Richard was getting on. He was sitting beneath one of the street lights and in its pool of light I could see that it had started to rain again. He made a very sad picture, huddled under his brolly in an oversized coat gazing abjectly across the water, hoping against hope that the otters would appear. Suddenly, his face lit up and he hissed at us.

In a flash Charlie had cast aside his supper, set the camera rolling and crawled into the corner of the deck. I was slightly slower and a little more reluctant, but remembering our great cause wrapped up my chips and dived onto the deck behind some oily, smelly, net-hoisting equipment. Simon had installed himself in the doorway to the cabin on a small camping seat, and continued to eat, oblivious to the dirty looks. He remained hunched over his packet of chips until Charlie whispered at him to wrap them up and quickly. We weren't sure whether otters would like the smell of fish and chips, and couldn't risk putting them off.

We waited. I was almost afraid to breathe in case the noise scared them away. The wind whistled around the tops of the boats and in the light I could see the rain had become quite heavy. I studied Richard up on the quay to see if he could see anything, but his face gave nothing away.

Minutes passed and nothing happened. I decided I should breathe and began to look around me. Once again I congratulated myself on my glamorous choice of career. Here I was, lying on the deck of a boat, covered in black oil from the machinery, everything imbued with the stench of fish. I was cramped as far into the corner of the deck as I could get, behind a heavy winch. I had hardly eaten any of my chips and was praying that no one could hear my stomach rumbling beneath the layers of thermals, fleeces and jackets and Mountain Toasties that had turned me into a small hippo. My hair was plastered to my face, my make-up had long since dissolved and snot ran from my nose to my mouth in an attractive rivulet that I could no longer feel because my face was so cold. But I felt alive.

In such a situation you'd think the hours would fly by, but they didn't.

'How much longer?' Simon was beginning to get restless on his three-legged camping seat. He looked like some kind of misplaced gnome, short with a pointy hat and some headphones, the sound mixer in his lap and the long boom mike angled from his body like a fishing pole.

'How long is a piece of string?' came the reply.

'Is this how you always make these kind of programmes?' Simon hadn't filmed any natural history programmes before.

'Yes.'

'What, you just sit and wait on the off-chance that an animal will turn up?'

'Yes.'

'That's daft that is.'

'Yes.'

The wind blew the rain in horizontal sheets. My feet grew numb, my legs and bum already were; I was going to have to move. Slowly I uncurled myself and peered over the edge of the boat. In spite of the wind, the water was very calm, barely a ripple, the lights of the town reflected back from the surface.

'Having a thoughtful moment?' Charlie was obviously bored too.

I took the breath to reply, but as I did, realised that there were ripples, two of them, perhaps two hundred yards away, making their way towards the three boats on the dock. But what if I was mistaken? It would be so embarrassing!

'I think ... otters!' I whispered, and dived for my oily corner.

Again we waited. Richard had spotted them too, but the expression on his face, wild and searching, indicated that he had lost sight of them. I started imagining what I would say when they showed up. How close would they get? Would I be able to express how unusual this behaviour was, how the otters here at Kyle were the only ones in the world known to have adapted in this way? Would I have the time? Would I have the presence of mind? Would my lips be able to form words despite the fact that they were numb?

My racing mind was interrupted by a strange rumbling noise, which disappeared almost immediately. Did otters purr? Was this a gap in my research? There it was again. What did it mean?

I tore my gaze from the deck to look at Charlie, who had one eye glued to the camera. He looked back at me with his free eye, but his face was strangely contorted into a grimace, and I couldn't decide whether he was furious or hysterical. Once again the strange rumbling. I frowned. Charlie indicated the cabin door with his head.

Our resident gnome had fallen asleep. His head lolled

forward onto his chest, and the incredible amount of padding he was wearing had wedged him into the doorway like a cork so he remained upright on his seat. From his chest emanated the deep rumbling of a snoring sound man.

I tried so hard not to laugh, but it was exactly that feeling you get when you're in church and the moment is so inappropriate that you just can't help it. The tears began to roll down my cheeks and my stomach ached from trying to control the giggles. Charlie soon gave up trying to hold it in too, and before we knew it we were hysterical and every grunt and snore from Simon made it worse.

We didn't see the otters again. It seems they weren't that impressed with the snoring and visited another boat that night. Simon missed out on the bar but got a great night's sleep. Charlie and I spent the rest of the trip laughing. And that's when my life began to change.

Chapter Two
TAKING THE PLUNGE

....

*In which Mole and Ratty find
a riverside residence*

Everything changed very rapidly and yet it felt as though it had been like this all my life. We met each other's families and friends, spent very little time apart, and before I knew it Charlie and Bill had moved into my London home with me and my cats Bert and George. This took a bit of getting used to for everyone else but to us seemed perfectly natural. We had fallen into each other's lives and that was how it stayed, just as though this was nothing new. It didn't feel unusual in any way; at the end of our Scottish trip we simply didn't part.

There is no better place to use cliché than when talking about love, and Charlie and I were like two peas from the same pod. We had the same passions: I was passionate about

Charlie and so was he. We both loved the natural world and got worked up about conservation issues, food and family. We loved to go for long walks, our favourite dogs were Border collies, open fires and whisky were important, as were long, lazy dinners with friends. There was only one area in which we differed, and that was our taste in music. I am more of a pop queen and if you can do aerobics to it I like it whereas Charlie is a Frank Zappa and Bob Dylan man, but at this stage in our relationship we were so in love that we even listened to each other's music with not a murmur of complaint.

Then there was television. Neither of us could remember not wanting to work in television. I had had the odd fantasy about becoming a vet and Charlie of becoming an author, but really we were both driven to make good TV. My kind of television had taken me to London and given me a great lifestyle. I would mix and mingle with rich and famous people, go to the odd première and was able to afford designer clothes. Charlie's kind of TV had based him in Bristol, given him a great knowledge of survival in rough conditions, earned him very little money and generally meant that all his clothes had holes in them or had gone mouldy in the rainforest. Both of us had travelled the world for many years and were reluctant to admit that we had a hankering to settle.

Given the way that nature works, it was also perfectly natural that I got pregnant very quickly, and this did take some getting used to. Only months before we had been single and now, suddenly, we were very much a couple and about to be a family. I think both families handled the 'surprise' very well, given how new the relationship was, but it took them all a while to adjust.

Yet for us, despite the shock there were no doubts. Life had suddenly become richer in every way. We were very happy together. In fact, in the tradition of true romance it simply felt as though it had always been that way. Eventually, however, despite the glow of love I had to face facts: trying to keep a

Border collie and a wildlife cameraman in London is just not fair. You see they need lots of fresh air and wild open spaces. They are really no good penned up in the house all day and get miserable pounding pavements and breathing in all those fumes. They need rabbits to chase and birds to watch and as a bare minimum require at least one good walk a day. As for Border collies? Well they need lots of fresh air and rabbits too.

Charlie and Bill did a great job of pretending that they were happy and fulfilled but it was obvious that London life did nothing to fill their souls. So we resolved to find another place to live. We knew that we needed to be in the West Country, mainly because the BBC Natural History Unit is based in Bristol.

We were in no rush and decided that it was important to find the perfect place. We spent happy hours talking about it and imagining exactly what that place would be like. On one bank holiday Monday we visited an old mill that was for sale in Wiltshire. It was an idyllic setting, deep in the countryside, and we were early.

As we sat waiting in the car I asked, 'How will we know if it's the one for us?'

'If a kingfisher flies past, then it's ours,' Charlie said. We looked out of every window but we didn't see a kingfisher and the house was way too big.

But that became the defining factor in our house search. Forget about conservatories, Agas, en-suite bathrooms – there was only one thing that mattered to Charlie: kingfishers flying past the windows.

Charlie was nine when he saw his first kingfisher, on a school outing in Bristol, and immediately fell in love. One flash of blue and he was hooked. He was also a member of the Young Ornithologists Club and used to spot them on YOC outings. But it wasn't until he was twelve that the infatuation developed into something deeper and, armed with his first camera, a

Nikkormat EL (what else?), he began to photograph them. He didn't actually get a decent picture until he was fifteen and to quote him, 'Even that was crap.' He was eighteen before he took a photo that he was pleased with.

Eventually, he learned how to cajole them onto the right perch, what they liked to eat, how and where to catch their favourite fish, where they nested, where the territorial boundaries were and what their young looked like in a nest. He is one of the few people to witness kingfishers fighting and trying to drown each other. He saw them mating and escaping from sparrowhawks, fishing and grooming themselves. He became familiar with their body language – all their different postures and what these meant. He watched the young learning to fly and fish, succeeding and failing. He saw every aspect of the kingfisher's life and this was just the start of a lifelong study. He became a man obsessed.

Charlie did all this on the Tipple, a small, unassuming West Country river. His school had arranged an outing to an adventure centre on the river, so that the children could draw the things around them, play in the adventure playground and enjoy nature walks. Charlie, however, was interested only in the fact that he had spotted a kingfisher. The river very quickly became the focus for all his spare time, and anyone who could offer him a lift from Bristol city centre to this haven was easy prey. If he couldn't get a lift he walked, miles and miles. Weekends, days off and days when he was meant to be at school were spent in wellies, hiding in bushes, taking photos of blue and orange birds.

Charlie and I decided to allow ourselves eighteen months to find our nirvana but in the end I found it in two. It was an accident, and happened while Charlie was away filming giant otters in Peru for *Wildlife on One*. Just like all the best things in life, it came along when I wasn't even looking.

It was late spring, I was working like a maniac and was

about four months pregnant. One night I returned home late to find a nice cream envelope lying on the mat in the hallway. The address was handwritten in pen and ink.

I am notorious for picking up post, putting it in a big pile and leaving it there. I don't even open it. I am usually so exhausted when I get home in the evenings that I just want to enjoy being there. It frustrates everyone so much that one friend, called Hilary, has taken to writing on her envelopes in big red letters 'Urgent!' or 'Open now!' so that I don't miss something important. Anyway this didn't look like business post to me; it seemed far more exciting, so I opened it straight away. Thick cream paper and a lovely typeface.

The address at the top was 'The Manor' – very posh but slightly confusing. I didn't come close to knowing anyone who lived in a manor. I read on. It seemed that Charlie had already begun some research.

> *Dear Philippa and Charlie*
>
> *Along the River Tipple there are perhaps 10–12 houses in total that would fit Charlie's thoughts of being near the Tipple to continue his lifelong studies.*
>
> *Of those perhaps 5 are sufficiently in the country and fall within your price bracket. In 8 years of living here I've not known one come up for sale – not that we scrutinise the local paper for those requirements that carefully. But lo and behold, here is one that would suit Charlie's thoughts.*
>
> *The agent Pritchards have had it a week. They have handed out 20+ brochures already. The specs sound interesting. The boiler system does not sound great. The rooms are all large but you may definitely need more than 2 bedrooms right away. Please see that it is attached to another house of similar size. The house is the next one south of us down the river, about 1 mile.*
>
> *One way or another, if you are serious about this step, it may be the best shot you get on the Tipple for a while.*
>
> *Call us if you'd like a little help or would like one of us to visit with you.*

Apologies for throwing you what may be a dilemma!
Yours,
Richard Horton

He had included an estate agent's brochure with a picture on the front.

There was a house, a white house by a river with a bridge. I opened the brochure cautiously. It felt heavy, not with paper or because something was inside it, but with portent. Glancing at my belly, I wasn't sure if I wanted any more of that, so I took my time. Phrases jumped off the page: 'charming mill cottage', 'rural setting', 'Californian-style kitchen' (what on earth was that?). Inside there were more pictures, the river and a waterfall! My heartbeat quickened. I tried to ignore it.

Something deep in my imagination began to stir, memories of childhood dreams. I had always loved the water and had fantasised about living on a river with a bridge to my front door. I pulled myself up quickly. I couldn't get too carried away. This was real life; there would be a price. What was the price? Actually, sod the price; it wouldn't do any harm to look. I took a deep breath and phoned Mr Horton.

He was charming. He had a very kind voice and was full of apologies that he might have overstepped the mark but anxious that this might be a perfect opportunity for us. He had known Charlie since he had first moved to the manor, where Charlie had pretty much come with the garden furniture as the person who turned up to film and take photos on the river.

It just so happened (another fortunate coincidence) that I was working in Wales the next day and my train from London went via Bristol. We arranged that if the estate agent could make an appointment I would hop off the train on the way back and meet Richard, who would come with me to see the house. I found it difficult to sleep that night.

The next day I made sure we finished work early. If only I always had that much control. The estate agent was in agreement and the appointment was set. By the end of the day I was growing increasingly impatient. It seemed an interminably long journey from the station to the Tipple valley and the baby in my tummy was trampolining on my bladder. Finally, after the taxi driver had lost and then recovered his sense of direction several times, we pulled into the driveway of the house. I was late and frazzled and desperate for the loo. It was drizzling and grey, a miserable day, but I barely even noticed the estate agent in his smart dark pin-striped suit as he shook my hand vigorously and showed me up the garden path.

Richard Horton was there, an unassuming but dignified man in his late forties. He looked in good shape. As I was to find out later, this was because his self-discipline gets him up and out of bed early every morning to run around the country lanes whatever the weather. His hair is greying and his mouth thin, his grey-blue eyes set in a face that somehow wouldn't have looked out of place in the Roman Senate, though he lacks the famous Roman nose. He is a naturally quiet man, but there is a shrewdness about his mouth and eyes, a certain cut of the jaw, which gives away his cleverness and determination. Always polite and yet very warm from the first moment you meet him, he loves his home and his wife Stephanie, and runs his own successful business, but is never happier than when he is up a ladder pruning trees, digging a pond or sharing a decent bottle of wine in front of the fire. I didn't know any of that then; I just knew I liked him but I am ashamed to say I said little. My attention was completely taken up by the house. It wasn't that it was picturesque. There were no roses round the door and it was plain, painted white with brown modern windows and a modern front door. The sky was grey behind it and, despite the time of year, the rain fell cold on my arms, making me shiver. But it was the river that made me catch my breath. And the path to the front door – fifty yards and then over a bridge, which was next to a waterfall.

The waterfall was really a weir, approximately fifty feet long and ten feet high, and the water trickled over it, filling the air with the constant sound of running water. The house had originally been one of the millworkers' cottages, and the mill itself was the next house down the river. Years ago the river would have flowed straight through the mill but not now. What did remain, though, was the sluice gate halfway across the bridge. This was fully operational and responsibility for its correct use remained with the people who lived in the millworkers' cottages.

I floated across the bridge – I don't remember my feet actually taking steps – and into the house. Almost immediately, I asked the first question that a pregnant woman asks upon entering a strange place: 'Where is the loo?' After all that trampolining I sat down in the bathroom with a sense of relief but something else as well: a strange feeling that I had arrived home. On the wall ahead of me was a mural of an Italian scene, beautifully painted, and on the ceiling above me fluffy white clouds tinged with pink raced across a blue sky, again all hand-painted. The current owners had put a lot of effort into the decorating. As I returned downstairs I had a strong image of us living here – a baby asleep upstairs and a sofa in front of the fire. I didn't need to see the rest of the house. It would need quite a bit of work to make it feel like ours, but it really was our home. At least I felt as though it was.

I wandered through the living room, gazing out of the windows. The rain was making big, round, spreading ripples in the millpond outside. I stepped into the hallway. It would be over a month before Charlie was due back, and only then if they had got all the shots. How could I make a decision about where we were going to live without asking him? Just as I was wondering what Charlie would think, a kingfisher flew right past the front door.

Richard saw it too and we laughed together about what it might mean. I had no doubts at all. I put in an offer. The next day it was accepted.

Several weeks later I had a rather crackly call from a satellite phone. Charlie was sitting on the banks of a river in the middle of the rainforest. I passed on the news that we had a new house. As I described it he knew exactly which house it was – I had forgotten that of course he knew every house on the river. His voice began to shake and I don't think it was a dodgy signal. You could hear the cheer without the aid of a satellite phone.

It was late May; Charlie had been due back from Peru for a week but there had been problems in the rainforest. Severe weather and lots of rain had meant that the landing strip where a small aircraft was due to land and pick him up had turned into a swamp. After waiting a few days he realised that there was going to be no change in the weather for the foreseeable future. So he had to find another way home. First of all he hired a boat and motored up the Madre Dios river for a day from Boca Manu Fitzcarraldo, but when it began to get dark the driver refused to go any further and Charlie found himself in the small village of Atalaya. So, along with his fifteen metal flight cases, Charlie had to hitch a lift in a truck which, after a three-hour drive and a charge of $100, dropped him off in a dark muddy hamlet with the promise that he would be picked up by a van. The van came after a couple of hours, and then it was a twelve-hour drive over the Andes to Cuzco. Then came a plane trip to Lima, and another few days waiting for a seat on a flight home to Heathrow.

At Heathrow, Bill and I were waiting. It was Bill's first trip to the airport and we found a quiet corner. A customs man offered him a drink and he was very grateful, showing his thanks with some tail helicoptering and a good deal of head inclining; the customs man was in love. The flight was delayed. I had made sure that our quiet corner was in good sight of a flight monitor with arrivals information on it, and every time the green letters came up to say it was delayed by another ten minutes my heart sank. It had been two months since we had last seen each other, and we had only spoken a couple of times, although I had had letters by the sackload and once a camera film for me to get developed which showed photos of the camp, the river and a heavily bearded Charlie carrying his camera through the forest. (Looking at that picture it wasn't hard to imagine where the yeti myth sprang from.)

Bill had a couple of slurps from his bowl of water and looked at me. I explained yet again why we were here and told him about the delay. He seemed satisfied and tried to settle between my feet and the chair but he was too big so simply sat on my feet and looked at me apologetically. I wondered what Charlie would look like. Would he still have the beard? What was he thinking, up above me now as he circled over west London? More to the point, what would he think of me? I had grown, significantly. When he had left I had barely a pound extra to show that I was pregnant; now I appeared to have a duvet stuffed under my jumper. And there was news to bombard him with: we had exchanged contracts on our dream home and had only a couple of weeks to wait before completion; we would be in for the summer. Also we were due to go and see it today, and were first invited for lunch with the neighbours.

The flight landed and then the awful wait; with all those cases to see through customs he could be an age. There is so much paperwork involved in travelling with filming equipment and one inexperienced customs person can hold you up for hours. But today we were lucky. Charlie soon came through the gate backwards pulling two trolleys of boxes which towered over even his six-foot frame. Bill's tail went wild. That black and white tail went round and round so fast that his backbone had trouble maintaining overall balance.

Charlie seemed very thin and when he turned to greet me his wild hair and dark-rimmed eyes revealed that he was a bit shell-shocked after his journey. But there was little time to rest; he had only twenty hours before we were both due on another flight across the Atlantic to go filming together in the southern United States and he was determined to make the most of his time in the country.

He was surprised and ecstatic to see Bill and they shared a joyful reunion on the grey plastic floor with much hugging and belly-tickling and strange half growling noises from both of them. Then it was my turn. I simply settled for a hug, the belly was too big for tickling and I could take or leave the growling.

It was of course wonderful to see him and have him close by again. We quickly offloaded the cases and kit, giving them to a BBC driver who had been entrusted with taking them on the last leg of their journey back to Bristol, and then before we knew it we were in the car and free and together. All the way down the M4 we talked. Charlie was barely able to keep track of where we were going as he told me about the six-day journey. As I drove I told him that the baby was now kicking, and the conversation dwelled a little on the giant river otters on the Amazon, but inevitably we quickly got onto the new house.

'It has been so easy. It feels like it was meant to be; the survey went through with no hitch – they just didn't have anything to compare it with so it took a little while. The building society have been super-efficient and I have found a lovely new solicitor who has really helped push things through. We are due to exchange next week.' Suddenly there was so much to say, and I realised that although there had, miraculously, been no stress involved in this house purchase, it was a joy to share it. It was such a relief to be able to sit and talk without the expensive minutes ticking by on a satellite phone.

'It is the white one right on the river, isn't it?' He was keen to know that he had been dreaming of the right place.

'It is exactly that one. Where the water falls over the weir just outside the window. We are going to see it now, today after lunch.'

I knew Charlie would be tired but would not want to turn

down our new neighbours' lunch invitation. They were Charlie's old friends Richard and Stephanie Horton.

The English sunshine was bright and clear, and as we travelled west into the countryside the many different shades of green surrounded us. Charlie just sat back in his seat for a while, his hand on Bill's head, his feet and legs squashed by the bulk of the Border collie in the footwell, and grinned.

I had not yet met Stephanie, who had been away when I had first viewed the house, and I was intrigued to know what she would be like. I was slightly apprehensive, as she had been in New York on business and I wondered if she was a real cut-and-thrust high-flyer. She and Richard would be the only people I would know in the area to start with and I really hoped we would hit it off.

I needn't have worried in the slightest. Stephanie met us at the back door to the manor as just one dazzling smile; she was delighted to see us and made such a fuss of Charlie and Bill. Dark, her thick hair is cut into a bob with a fringe, her skin blooms with health and her eyes twinkle merrily. She doesn't get up at dawn to pound the country lanes but prefers to luxuriate in bed with her Jack Russell curled up beside her and the smell of coffee made by Richard slowly seducing her into wakefulness. As a consequence she calls herself fat, but she is just voluptuous. From that day to this I don't think I have ever seen her low.

I quickly realised that Stephanie is always the first person to put herself down. Would there be enough food? Should she have made more of an effort? Would we really be comfortable enough in the kitchen? But of course the kitchen was filled with lovely food, bread and cheeses and quiches, all the things that Charlie had missed desperately. We drank wine round the kitchen table. Richard quietly nodded and smiled and squeezed Stephanie's hand to reassure her that she and lunch were perfect. After recounting in detail his incredible journey – along with hilarious impersonations of his guide – Charlie had a longed-for shower and Richard and Stephanie filled me in on all the local gossip. As I listened I realised how comfortable I felt in this house and hoped that our new home would feel the same.

The manor is set in lovely flat grounds dotted with all sorts of trees and bordered on one side by the river. There is a cottage occupied by Stephanie's parents and all the buildings are of old Somerset stone. What really caught my eye on that first visit were the gardens, perfectly green, simple and symmetrical. A long lawn retreats from the kitchen window for at least five hundred yards. It is impressively even with wide, traditional herbaceous borders running down either side filled with all the old favourites, pinks, peonies and hollyhocks. Roses and wisteria climb the walls of the garden and the house.

A rectangular stone pond sits in the centre of the lawn about two thirds of the way down. There are no fish left because of the local heron population but there is a fountain whose gentle sound bounces off the walls at you as you wander. Two yew trees stretch up into the sky about halfway down the lawn, framing the pond perfectly, and in the distance, where the lawn ends, another ancient yew reaches out in many directions towards the countryside beyond. Closer to the house an old apple tree peers in at the kitchen window and a flagstone patio provides a place to sit. The garden is enclosed by ancient walls, all except for the very far end, where the view is open.

After lunch, we drove the short distance to our future home. The owners, Walter and Hilary, were there this time and were very welcoming. It looked even better than I remembered it; by now the garden had changed and bloomed with the season, and skeletal spring foliage had been replaced by lush green and bright flowers.

We were ushered through the door by a booming Walter, who wasn't old but was more than ready to grasp retirement by the throat and take it to Australia to start a new life. The landlord and owner of several local pubs he knew how to welcome people; he was smart and scrubbed in a perfectly ironed shirt and smelled attractively of aftershave. Hilary was a little glamorously turned out for my idea of country life, her hair styled and her make-up on. She greeted us warmly and went back to stirring some soup while Walter assured us that we had made the right decision. He showed us round the house pointing out all the features he was most proud of – the way the music from the stereo was piped into every room, the twin hot water tanks, one for his bathroom and one for Hilary's so that the husband need never suffer as a result of his wife's extravagance with hot water, the corian surfaces in the bathroom, power showers and jacuzzi baths -and then back to the kitchen for coffee. Walter had lavished a good deal of time and money on the house, while Hilary had kept it beautifully clean and made sure that the extravagance still balanced in the books. When she could get a word in edgeways she told us about the garden and the man who delivered the coal and what they would be taking and asked what we would like them to leave.

While the kettle boiled and the coffee brewed, Charlie sat quietly in the kitchen with its wonderful views over the river, entranced by the kingfishers coming and going and the wagtails dancing on the waterfall. Two horses gazed at us over the gate on the other side of the lane, and I gazed back at them with a big grin on my face. It was sunny; on my previous visit it had been raining and today it looked one hundred per cent better. I could hardly believe that this little piece of paradise was to be my home. Although we were pleased to have them, the jacuzzi and power showers were bonuses; had it been falling down around our ears this house would still have been our home. I was too overwhelmed to do anything practical like measure for curtains or furniture, and all Walter and Hilary's instructions on how to operate the sluice gate when the river rose and what to do with the temperamental boiler seemed to flow straight over my head. They reminded us that they wouldn't be around for much longer to ask because they were emigrating to Australia where they were having another dream home built – the furniture was being shipped out in just a couple of weeks – but despite their sense of urgency we just grinned, captivated by the view. They could obviously tell because they promised to write everything down.

Charlie and I spent the next six weeks working hard and travelling a lot. After our trip to the southern states of the USA I went to Australia, and we finished up with a visit to one of Charlie's favourite places, Shetland. He was working on a short film about the otters of Shetland for the Natural History Unit but afterwards we took a couple of days off so that I could finally see some wild otters for myself. After all, the Skye trip had not been exactly successful in terms of otter sightings. This time Charlie was determined I would get to see some.

In Shetland in summer it just doesn't get dark and at midnight we could be otter watching on the beach surrounded by a load of sheep, who were trying to sleep and obviously thought we were very odd judging from the strange glances they gave us.

We had been on the lookout for one particular otter Charlie had filmed before, a small female. This female had cubs and so we were careful not to get too close to where we thought they might be on the beach. We first saw her in the water, fishing. She had grabbed a crab, a huge crab at least the size of her head, which was not happy to relinquish his life without a fight. However, he would make great food for the otter's cubs. After what looked like more than a minor tussle in the water the otter seemed to have him under control but the strange twelve-legged beast in my binoculars appeared to have grown; during the grappling the crab had grabbed hold of an unfortunate jellyfish which was obviously in the wrong place at the wrong time and which the crab would not relinquish either. It seemed that if the crab was going to die, then he was taking the jellyfish with him.

Whether it was because this was a particularly difficult crab and she was taking no chances or because by now her cubs would be getting hungry we didn't know, but suddenly the otter turned towards the beach and began speeding towards us.

I was near the top of the beach and Charlie was close to the water with his camera; we had been moving along the beach each time the otter dived, taking advantage of the time she was under the water to get nearer without distracting her. Each dive lasted around thirty seconds. I, of course, was not very nimble on the rocky beach, being rather large and unbalanced, and had not been able to move as fast as Charlie, so was lagging behind. Now that she was swimming along the surface of the water, wrestling with a crab and a jellyfish, there was no way the otter would dive again and risk losing them. We had no time to move and she was heading straight for us; it seemed we had misjudged the position of the holt and those precious cubs.

I tried not to breathe too heavily and Charlie dropped down, all six foot-odd pressed flat against the beach in the shallow water. He looked at me urgently; this could be a great chance for him to get the shot he was after. I presumed his eye-to-eye communication meant that I was to do the same thing and drop down flat to the beach like some kind of SAS soldier on a mission. Well, this was not as easy as he thought, in fact there wasn't a hope in hell on account of the fact that I had all the weight of a standard-issue army rucksack on my front. I could hardly bend down to the bottom shelf at Sainsbury's, let alone drop to the beach in seconds. I sighed and looked at the otter fast approaching, then saw the anxious pleading in Charlie's eyes, apologised to the baby and did my best.

I got down onto the wet, slimy, fishy-smelling rocks surprisingly quickly, but there was no way that I could lie on my stomach as it was like a big round ball. So there I was on my hands and knees. I had waterproof trousers and a big red puffer jacket on so I really needed to try to flatten myself against the rocks, to melt into them and not be seen. But it was no good; the otter was getting closer and closer to Charlie, who was looking at me aghast. I tried to lie on my back but the contortions on Charlie's face seemed to indicate that that was no good either. My belly must have been sticking up like a huge red hill on the beach and I could no longer see anything. I only had one more option: I would have to lie on my side. But this wasn't easy either; it was as though – and I hate to use this term – I had turned into a beached whale. My short limbs sticking out of the puffer jacket and trousers were, and there is no more dignified description, flailing. I lay there loudly rustling and flailing like a gigantic chrysalis as the otter approached. Charlie looked as though he might cry.

But just in time I made it. By some miracle of physics the belly lurched to one side and then took the rest of me with it. And there I was, as flat as I could be, with my hair and ear in a sandy puddle, watching, at otter level, our female leave the water.

It was as though we didn't exist. She left the water not two feet from Charlie, almost too close for him to focus. And then she began the long task of dragging the crab, with his jellyfish hostage, all the way up the beach over the rocks. She was heading straight for me the whole time and, I may as well confess, this was one red beached whale that was breathing heavily and loudly due to general exertion and much flailing. But to watch her you would never have known I was there. Charlie was in shock. Had he reached out his arm he would have been able to save the jellyfish, that was how close he had been.

Her concentration was intense. We didn't know if it was because of the cubs or because she had to work so hard with the wretched troublemaker crab that she had caught, but we could have been doing the cancan for all she cared.

And then it was my turn for a close encounter. Up the beach she lumbered, and if she had been human I am sure she would have been cursing the crab. She crossed in front of me about four feet from my head. I could see each drop of seawater on her coat, each whisker and finally the expression on her face – dogged determination. Then she disappeared behind a rock just a few feet from where I was lying and into her cosy holt in the earth bank which had formed over the years at the top of the stony beach. There, presumably, a couple of cubs were eagerly anticipating shellfish for lunch.

Charlie made his way up the beach, pulled me to my feet (not an easy task) and we moved back to the car as quickly as we could. We didn't want to be there when she came back in case our presence made her think about moving the cubs or distracted her from valuable fishing time. It had been and still is my closest sighting of an otter, so it was a rare treat. However, I shall always remember it for the comedy of the poor jellyfish and the tenacious crab.

Chapter Three
THE WATERS BREAK

New neighbours and new babies

In the middle of July we moved. I was unceremoniously heaved into the cab of a rickety hired van, which took two men and some effort since I was by now six months pregnant and big. We rolled down the M4 and proceeded to carry all our possessions over the bridge into our new home. I still quake at the memory of my favourite furniture being marched precariously across the bridge and wobbling a little over the weir.

We settled in very quickly and so did the dogs; Bill's girlfriend had come to live with us. Honey belongs to my in-laws-to-be and is another Border collie. She is very pretty, small for a Border collie, with a little foxy-shaped face, bright eyes and perky ears. She has so much feathering down her back legs that she looks as though she is wearing some kind of crimped fur bustle. She is named well; you couldn't wish for a dog with a sweeter nature. She is very affectionate and soft and like every Border collie I know has boundless energy and enthusiasm. She would be company for the ageing Bill, a great guard dog and above all would have a fine life with us running around outside all day in the huge garden. It seemed like a great idea.

I had never seen Bill so full of bounce, his tail was doing helicopters from the moment we arrived, spinning round and round as though there were no spine to hold it in place. He had a huge new garden, paddocks full of rabbits to chase and a girlfriend to share it all with. There was just one drawback: Charlie couldn't and still can't bear Honey; she is just too enthusiastic for him. She is scatty and not very disciplined. She will sit when you tell her to but only if she is listening. She is eager to please but completely unsure how to do this, and her mad energy spills over into chaos. Charlie prefers a dog like Bill, who was trained to exacting standards and will do whatever you ask him to. Bill is also bright and so is easy to be with. Other people pick up on this very quickly because when they talk to him he looks directly at them and cocks his head from side to side to indicate that he can understand every word. He is a good listener and universally popular as a consequence.

The fact that he can learn new things so fast also comes in handy. For example, when you can't be bothered to get off the sofa and close the door, he will nearly always oblige if you ask him several times. He spent much of his early years filming in Scotland with Charlie, where they would stalk otters together and he had to learn to be inconspicuous and stick close to Charlie's side. They would canoe to islands off the coast and have all sorts of heroic adventures, all of which have been related to me over and again on cosy nights in by the fire. Adventures like the time Bill nearly died on Skye after fighting a big male golden retriever. Bill was losing, and instead of inflicting any injury at all on the other dog, he managed to bite a hole clean through his own tongue. This would not stop bleeding so Bill had to be removed from the fight and taken to the vets. He filled the Land Rover with blood and covered the vet's front path and the surgery itself (by all accounts this is no exaggeration) and had to have two stitches under a general anaesthetic before the flow was finally staunched. Heroic tales, adventurous days – just one man and his dog.

Bill is not perfect; he has his faults. He is especially grumpy with children (we never let him close) and other male dogs, particularly now that he is getting old and stiff; he smells; he is frightened of loud aeroplanes and any kind of banging; he is slightly neurotic. Every summer if he feels he is not getting enough attention he comes down with a mystery illness. It is bad enough that we have to steel ourselves for the inevitable, but he ends up costing us a fortune and stretching the vet's powers of deduction. Then, suddenly, for no reason anyone can fathom, he is fine again. His tongue is also too long for his mouth and permanently sticks out of the end. When he has been asleep for a long time it goes all dry and crinkly and picks up bits off the carpet, then it stays crimped and fluff-covered when he wakes up. Having the end of his tongue sticking out of his mouth does him no favours when he is trying to look intelligent either.

Bill, like many dogs, is obsessed with marking his territory, but he will often take it too far and once did it against the grand piano on the cream carpet in my producer's house. However, Charlie is blind to his faults; in his eyes Bill is perfect. But then in Bill's eyes Charlie is perfect, so their relationship is flawless.

Honey and I had something in common too: we were both considered way down the pecking order. I loved her. So did Bill, passionately on the patio while I was unpacking the plates in the kitchen. The arthritis which normally seemed to plague him seemed miraculously to clear up for thirty minutes that afternoon.

Charlie and I passed the next couple of months launching an attack on woodchip wallpaper. It was everywhere, and when you scraped at it half the plaster underneath came off. The cottage had beautiful ceiling beams that had all been painted white, so long-legged Charlie embarked on sanding them down. After a week his arms ached so much that he resorted to using an angle-grinder. In no time at all the beams were stripped bare, although the ferocious nature of their stripping left them looking a little more rustic than we had intended.

We mooched around shops looking for furniture and slowly got to know our neighbours. Closest to us was Delia. She lived just next door, over our bridge, across a lawn and back over the river on the other side of her bridge. Although well over the age of retirement she was slim and pretty and very fit. Her hair was white, yet yoga and plenty of walking meant that her posture revealed no signs of ageing. Far from being a little old lady she seemed vibrant and strong. She wore little make-up, just enough to look her best and add a glow to her cheeks; her clothes were comfortable but still coordinated as was everything in her cottage. I constantly found myself wondering how beautiful she must have been in her youth. She welcomed us on the first day with a huge hug on her bridge, and from that moment we were destined to be friends.

'It's lovely to have two young people moving in, I'm so delighted. Anything you need just ask, and before you do anything, you must come in for a coffee.'

She was a wonderful neighbour. There was nothing she wouldn't do to help us, constantly calling in with Charlie's favourite cheesecake or gardening advice, growing enough seedlings to share and offering company whenever I was lonely. There was nothing in Delia to suggest that she might even consider giving in to either old age or gravity. Her cottage always felt sunny since it was full of her favourite colour, yellow, and had yellow shutters on the outside. The yellow window boxes on every sill cascaded plants down the outside of the house and her garden was a riot of colour and shapes and worthy of an award at Chelsea. Flowers burst out of every corner and shouted from all the beds. Delia was frequently to be found on her hands and knees weeding, even on rainy days. Everything was neat and tidy, from the lawn to the greenhouse, and she was followed everywhere by her adoring shaggy dog, Tansy.

Delia had only recently lost her husband and although you could see the grief in her eyes when we talked about him, she kept herself surrounded by people and seemed very brave. We spent many happy evenings there having supper and sharing stories. The first time we were invited round we crossed the bridge to her house eyeing the steamy kitchen window shining yellow out onto the river and dreaming of the poached salmon and new potatoes we knew were waiting inside. I was musing on how I was beginning not to miss London restaurants and reaching for the knocker when the yellow stable door was flung open and a rosy-faced grinning Delia announced to us and the river, 'Hello, I seem to be a bit squiffy. I've made a good start on the wine.'

She then swayed and sashayed around the kitchen, preparing the nicest supper and making us laugh. I think she was glad of the company, although she was by no means lonely, and she was thrilled at the thought that there would soon be a baby to cuddle just next door. After that we spent many happy evenings at Delia's house 'getting squiffy'.

The cats also made the move from town to country, in my best friend Tina's car, with me in the passenger seat and Tina at the wheel. Bert and George, both female moggies, one black and one black and white, had been my constant companions for nearly ten years, and had lived in London all that time. As a consequence, they didn't bother with outside life much but preferred to stay in and socialise. They weren't too impressed with the thought of rural life. It was the noisiest journey I have ever endured. They treated Tina and me to a two-hour rant on the joys of London all the way down the M4, and when they began to feel that we weren't taking enough notice of them they raised the pitch in order to beseech the people in cars on the other side of the motorway to please return them east to the city they loved. We couldn't hear the radio, we couldn't hear each other; there was nothing to do but sit and listen, each howl echoing around our brains. There was a short reprieve when we had to make an emergency stop and the cats were stunned into silence for about ten seconds when they found themselves on the dashboard.

When we finally arrived they were not too thrilled to find that they would be sharing their rural idyll with the huge black and white hairy thing that nicked their food and cluttered up the area in front of the fire, but then Charlie was never a great cat fan and nor was Bill. Dogs, cats and Charlie soon settled into a relatively peaceful harmony; the cats took over upstairs and Bill and Honey settled in downstairs. If Bill ever dared to try to climb up into cat territory he would be ambushed. We would find him whining, cut off at the pass, unable to go backwards or forwards, a hissing cat on either side, and he would have to be rescued.

Aside from the woodchip wallpaper, the house itself needed very little work – just a little decorating and a new fireplace to make us feel like we had made our mark. The difference was the pace of life and, of course, the river. It is ever-changing. As you move through the house it is impossible not to find yourself staring out of the window in each room. Ten minutes here and ten minutes there disappear from your day, from one minute to the next the colour on the water changes with the light or a duck swims over the surface, leaving ripples in its wake. There is always something to see, always something to distract you and hold you and mesmerise you.

Much of the time we entertained. As my waistline increased with the baby, my workload declined and the house was constantly full of visitors. Everyone seemed to want to come to this place. Our closest friends visited and then simply stayed. The sun shone brightly and the balmy days beside the river were better than being in the south of France. We lounged and chatted and lazed and ate barbecues. People frequently called in sick to work in London and extended their stay by a few days.

The river is inspiring in different ways to different people. Some are happy to sit and stare, others get the urge to play on it or in it. We did persuade one friend to swim. The water is very deep just outside the house and at least ten feet in the millpond just in front of the patio. Here the water is cold and dark. I mean really cold. On the hottest days I would sit on the steps leading down to the river and dunk my poor pregnant, swollen feet in the water, but after just a few minutes I would have them out again; any longer and they would go blue. This particular friend had only had a couple of glasses of Pimm's when we began our campaign to get him to swim under the bridge. Guy was particularly athletic, frequently cycling from London to Cardiff for fun and therefore quite keen to prove himself. We preyed on this mercilessly. It didn't take long before he had his kit off and was diving off the patio in his trunks, coming up gasping as the cold water forced the air out of his lungs. He managed a few lengths and went under the bridge a couple of times but was back out and lying on the warm flagstones of the patio pretty quickly and didn't ever mention going in again.

The kingfishers are the real stars, and all our guests are subjected to a crash course in kingfisher watching. All day the birds come and go, zipping up and down the river on important business. They tend to be feeding chicks through the summer and spend most of their time diving into the huge shoals of minnows in the water. The bridge is a favourite fishing spot and a great place to see them at their best; the railings are white and show up the famous kingfisher colours a treat. Everything stops when a kingfisher arrives on the bridge, and no one moves for fear of disturbing it. Bob-bob-bob, a peculiar dance up and down, gazing into the water and then the dive – fast, perfect and lethal. The bird perches on the bridge with its prize, a minnow, stunned and wriggling slowly in the long, pointed beak. The kingfisher twists its head to one side and thwack! The fish is killed, brained against the railing. Thwack again on the other side, just to make sure. And then the bird is gone, carrying the prize to its nest, and we all move again.

We found that even people who turned up claiming they had no interest in birds at all quickly became captivated by the kingfisher. The spectacular colours, the precision dives, the skill at manipulating fish with the spear-shaped beak, all seduce even the most uninterested person. There is no other bird like it, and rarely can you get close enough to see all these things in the flesh.

Our athletic friend Guy was particularly moved by the beauty of the bird, so much so that he was inspired to lie still for hours on end in order to capture it on film. It seemed that he had a secret yen to be a wildlife photographer. Charlie's job is one that on the surface looks so easy. At times even I have been sure that all it takes is a little patience and you too could get award-winning photographs.

One morning the sun woke me up, streaming through a gap in the curtains at 5.30, but I wasn't the first one up and about. I looked out of the window at the sunlight on the river and the birds going about their business and then realised that below me on the patio was the prostrate figure of Guy. He was propped up behind a couple of flowerpots with a few cushions and was holding himself completely still. His neck must have been killing him, but to his credit he didn't move at all. Clutched in his hand, balancing on one of the flowerpots was the carefully selected tool with which he would take his picture, a small Olympus automatic everything camera. It didn't even have a zoom lens.

I woke Charlie up and we watched Guy lie in wait for the kingfisher for hours, first from our bedroom window and then from the kitchen. The kingfisher visited several times. It sat on the bridge about fifteen feet away from the small row of flowerpots wondering why there was a man crouching behind them. We could tell when it arrived by the flashes coming from the patio.

Eventually the smell of bacon coming from the kitchen was too much even for the most dedicated photographer. Guy managed to force himself away from his work and staggered in for breakfast stiff and sore from lying on the York stone to discuss proudly with Charlie the variety of shots he had got.

When he got them developed we only managed around forty-five seconds without laughing. We did try not to, but it was impossible to rave over the small dots of orange and blue against a distant white railing which were all that could be seen of the halcyon bird, and even then you had to look closely. Our mouths twitched and twisted and our eyes filled with tears till, after a few minutes, we were howling. The further we got through the huge pile the more we howled and pointed and wept and pointed some more. When you could actually make out the kingfisher it was out of focus but many photos were simply of our bridge, occasionally with a blue blob just about to leave the frame of the picture. Guy's disappointment was evident; those, he explained to our delight, were meant to be action shots of the bird in flight. I laughed so much I thought I would give birth. Guy was a gentleman, magnanimous in defeat, and ever since he has been extremely generous in his praise of Charlie's skill.

That summer seemed to last for ever. One balmy evening when we had the place to ourselves we were eating dinner and chatting in the kitchen, while gazing out of the windows at the river. Walter and Hilary had had the brilliant idea of illuminating the river so that you could enjoy the view just as much at night. We were lazily discussing something that had been brewing in the backs of our minds for some time – making a film about the river. This river, our river, was only small but very special. To us it had everything that a British river should have and was the perfect example of a river ecosystem. Well, almost perfect. We had often joked that it lacked only one thing: otters.

Sadly, although they were the top natural predator on English rivers, otters had almost disappeared from this country and were now to be found in only a few places. Pollution and hunting had virtually wiped them out in the 1970s, even though their numbers had seemed healthy as recently as the early 1950s. The decline seems to have been due to organo-chlorines. Chemicals such as dieldrin found their way into waterways and then into the food chain. At the top of the food chain, otters suffered; their reproduction was affected and, when the pollution reached high levels, they died. They came very close to extinction.

I can still hardly believe what happened next that night. Out of the water, under the floodlight, not five metres away from where we were sitting, popped a brown furry head. Two button eyes studied us from above a square, whiskered nose, we stared back and then, plop, it was gone.

'Fu ... Fu ... Fu ...! An otter!' Charlie managed to stammer. It's very handy having an expert sitting next to you at times like this, because he can verify the sighting even if it isn't in a particularly scientific manner. After he had finished exclaiming Charlie ran around the kitchen for a while in the style of a headless chicken. I would have done this too but was so heavily pregnant that the most I could manage was to stand up.

What, you may ask, was all the fuss about? Well, we had just sighted a species that was not known to live in the area, and one of the rarest mammals in Britain. Even when otter populations were healthy it was highly unusual actually to catch sight of one. In Britain they are nocturnal creatures and extremely shy, and when they dive they are out of view for minutes at a time. Charlie had worked on this river for nearly fifteen years and had never seen a single sign of one. He had spent much of his working life filming otters in Scotland and Shetland and here was one popping his head out of the water to look at us in our own kitchen! Our otter obviously had a taste for loud rock music; we had the Rolling Stones' 'You Can't Always Get What You Want' playing full blast, the doors and windows open and the lights on, and we still saw him. Now, perhaps, you can understand why we were so excited.

The otter returned three times that evening, and because of the lights on the clear water, the final time we could see his whole body treading water underneath the surface as he gazed up at the house before swimming off. Where would he go that night? What would he eat? Would he meet up with other otters? Was he just passing through or would he return? As I sat in the dog basket (more comfortable than the floor at this stage of pregnancy) watching the ripples he had left, I knew two things: that we were meant to live here, and that this would make a great film.

Otters were returning to rivers they had previously abandoned and here was one just a few miles from Bristol. What a story! Now, surely the BBC would be enthusiastic about us making a film about our river. But how would we get it commissioned? Even for two people with lots of experience, the television business is a strange one.

Up until now it had all been very easy; somebody else had come up with the ideas and chaperoned them through the system, been responsible for the budget and the quality of the programme and how many people watched it. But with your own project comes a certain amount of responsibility and you want to do it your way. We were both ready for this step, and knew it. The question was, did anyone else? Would we be given the break we needed to make our first film? Whether or not we had the ability to do it never crossed our minds.

In the meantime we remained fairly busy. As my pregnancy developed so did my nesting instincts. I cleaned like a woman possessed until the house sparkled in the summer sun. We wallpapered, painted and cleaned the days away. Blackberries appeared in the hedgerows all around the house and apples on the trees, so I made jam like a demon. I hadn't quite got the hang of the setting temperature though. It should be simple: pour a little of the boiling mixture onto a saucer and when it is ready to set, if you push the surface, it will wrinkle. Mine never wrinkled; half an hour went by, saucers of jam littered the kitchen but not one would wrinkle. Eventually I gave up and poured the jam into jars, added pretty labels and sat looking at them and feeling proud of myself. I gave everyone I knew a jar including Delia and a week or so later we came to try our own. It seemed it had reached the setting point after all and in fact gone a long way beyond it. There was no chance of removing the jam from the jar without a pneumatic drill. Charlie smiled as he chiselled away at the ungiving surface as if this was something you did every morning at the breakfast table, and even as he used the hammer on the back end of the knife he kept smiling and murmuring encouragements like, 'What the bloody hell did you do to this stuff?' and, 'I'm sure it will *taste* lovely.' Meanwhile, all I could think of was that at breakfast tables up and down the lane hammers were being called into action while people smiled politely. When I made my red-faced apologies to Delia she said that she had warmed hers up and put it on ice cream, which made me feel much better. She then offered to get me a proper thermometer next time she was in John Lewis.

I wasn't put off though. I should have been; Charlie would have been delighted if I had ended my preserving career right there, but chutney of every variety flowed from the kitchen like a river. I found uses for all our surplus fruit, tomatoes, apples, even pears; including obscure recipes like 'orchard butter', a kind of midway point between chutney and jam which involved piles of pears and apples all over the kitchen. Mrs Beeton had never been so well-thumbed, the glories of the countryside never so plundered. The wine rack was full of white wine vinegar and cider vinegar, granulated sugar covered every surface and old jam jars fell out of every cupboard, the sound of swearing spilled from the kitchen and drowned out the birdsong. Cooking and cleaning were things I could do – theoretically. Move over Nigella Lawson; the domestic goddess slot was about to be refilled. Apart from a few outstanding stories, I had almost finished work and, embarrassed by my blue whale proportions, spent very little time away from the house. One day I caught sight of myself side on, and realised that rolls of flesh had appeared on my back. I remembered my slender, previous size-eight shape and nearly sobbed.

One of the strange things about being pregnant is that everyone feels they have the right to comment on your appearance and size, as if suddenly such remarks have no effect on you at all. People say things they would never dream of saying to you in a normal state. There is almost an assumption that as the pregnancy takes over and you have little control over the shape of your body, you also become divorced from any feelings about it. In fact, as every pregnant woman knows, when you are trying to grow a baby, undergoing all sorts of peculiar physical changes in the process and worrying whether everything will be OK, you are at your most vulnerable. Your body is doing its own thing and you have no control; you are proud but also a little frightened.

If anyone looked as though they might be about to make a comment, Charlie held his breath, knowing he would have to deal with the hysterical consequences. There is nothing more powerfully passionate than a pregnant woman.

My working life mirrored my personal life as the pregnant animals we were filming for a programme called *Making Animal Babies* grew larger and larger too. My favourite was a giraffe called Biffa who resided at Marwell Zoological Gardens in Hampshire and who kept giving us false alarms. The trouble was it was hard to be sure exactly when she was due; the gestation period in giraffes can be anything from 395 to 425 days, so very often we would race from Bristol to Marwell to find that she just had hiccups. Each time, Bill her keeper would phone with the same message: 'It's Biffa, she's acting very strangely.' The first few times, we'd get our hopes up and I'd cancel all forthcoming social arrangements and fill the car up with petrol, but after that we knew better. In the end I told Bill that acting strangely during pregnancy is perfectly normal. I knew.

We grew so desperate for any idea of a due date that we enlisted the help of a cranial osteopath, an expert who works with and detects the body's rhythms through the head and back. They treat problems stemming from bad posture or injuries and are very good in pregnancy when posture is changing all the time. Although it is easy to be sceptical, my own experiences have been nothing short of miraculous and I continued to see a cranial osteopath throughout my pregnancy. He stopped headaches, pelvic aches, stomach pains, and improved my well-being no end. I am a believer and I had no reason to suppose that it wouldn't work for Biffa.

So, once again we all trooped down to Marwell, this time with Trevor the cranial osteopath in tow, and once again they were very welcoming. We invaded the keeper's room at the top of the giraffe house, out of the way of the public and sat drinking coffee and tea under the watchful eyes of topless models stuck up alongside an assortment of stunning wildlife shots taken from an altogether different sort of calendar. There was an important problem to discuss before we started filming. We had the cranial osteopath, the keeper and the pregnant giraffe all assembled. The only difficulty would be gaining access to this particular patient's cranium. As is often the case with animals, we resorted to bribery and, as is often the case, it worked.

We established ourselves next to Biffa, and the camera began to roll. I held a bucket of carrots and every time that head swooped down from its great height Trevor had a quick feel. It worked like a dream, and the verdict was swift and decisive.

'She's definitely pregnant.'

And, as if in response to the treatment, little hooves inside Biffa's tummy began to move quite violently, as if the baby was having a rough and tumble. Then came the information we'd been waiting for: 'It will probably be Sunday. I can't say definitely, but probably. It could be *next* Sunday but I reckon it's a Sunday.'

We were so desperate for any kind of clue that we leaped on this nugget of information like a pack of hyenas, laughing and optimistically reorganising our bank holiday weekend socialising to accommodate some filming.

Of course she didn't drop that Sunday or the next. In fact when she did it was on a Tuesday evening. I know this because the call came halfway through *EastEnders*. Just when I was praying that Pauline would accept her first proposal of marriage since Arthur's death and live happily ever after, Dale the producer rang.

'Bill's been on the phone.'

'Really?' I replied, all attention still focused on Pauline, who wasn't reacting very well to the proposal.

'How soon could you be at Marwell? Do you think you could make it tonight?' Knowing that I was heavily pregnant, Dale was fully aware of how tired I was beginning to get and was desperately trying to be considerate.

'Why, is she "acting strangely"?' I joked sarcastically, frowning in my attempt to follow why Pauline was so upset. It seemed that she really wasn't open to Jeff's offer of lifelong happiness at all. He had obviously misread all the signals, but how? Was Pauline still in love with long-dead allotment-addicted Arthur, even after all these years? My heartstrings were warming up for a symphony – surely Pauline would see that it was time to move on.

Dale's voice cut through my musing.

'The feet are poking out.'

I was on my way. Off we zipped down to Marwell in record time, poor Charlie drafted in as high-speed chauffeur and stills photographer when he had planned Thai prawn curry in front of the TV. Sadly I missed the birth, although we did get it on camera for the programme.

Bruce, the father, had apparently got his head in the way when the waters broke so his experience of the birth wasn't as positive as he might have hoped. About an hour after the first signs of the baby, the six-footer dropped to earth with an almighty crash. Bill recorded it on a small camera and it was shocking to watch the pictures.

The baby fell at least seven feet to the ground. After landing it lay motionless, still in the birth sac, the seconds ticking slowly by. There was no movement. How could such a new, fragile being survive such a fall? Then just when you began to fear the worst the ribcage moved as it began to breathe. Biffa recovered from the shock and began to clean away the mess and massage her new treasure to life. The camera just ran and ran throughout with Bill doing a choked commentary. He was no professional and the lighting was appalling, it wasn't edited and had no music but it was one of the most amazing pieces of television I had ever seen. The long wait had been worth it. Some people would have been speechless with relief but fortunately for us the intensity of emotion seemed to act as a verbal laxative: just as though he were the proud father, Bill could not stop talking.

Only a short while later our very own Bill became a proud father, although he showed little or no interest in his offspring. While we were in London and she was staying with the in-laws, Honey gave birth to seven puppies. We dashed over to see her and them. Honey was installed in the warmth of the laundry room in a lovely large bed. The puppies were tiny, still just hours old and squeaking and blind and wriggly and warm. She had every reason to be proud. She let me pick one up for a short time although she kept close enough to touch it with her nose, and somehow with one week to go until I was due I felt we shared an understanding.

Just fourteen days later I was to find out what labour was like for myself. The birth went without a hitch, although a human infant, at a mere seven pounds and two ounces, was enough to make my eyes water. I was relieved not to be a giraffe. After nearly a year of watching and filming animals making their own babies I finally got to experience the miracle of mother-hood for myself. Charlie and I spent days gazing at our offspring and feeling proud of ourselves, and two weeks after his birth in London we brought Fred home to the river, which he took very little notice of.

Learning to feed, change, soothe and bathe Fred and trying to work out when we were meant to get some sleep distracted us from our film idea for a little while. Like every set of new parents it all took a lot of getting used to. Antenatal classes don't seem to venture beyond the point when the baby pops out, although there is some mention of the placenta, which no one likes to dwell on for too long. What antenatal doesn't cover is the mystery of colic – the fact that a new baby may cry in the middle of the night for ages and you will not always know why. Like every set of new parents we developed a survival strategy: in those early weeks of sleepless nights Fred was taken on badger-watching trips up and down the lane in the car to get him to sleep. Having left with a screaming banshee, Charlie would tiptoe back into the bedroom with a cherub straight from heaven and would gently place him back in bed. Some nights it would work but other nights the banshee would wake again and they would leave for another hour's badger watching.

Slowly over several months, it dawned on us that there was no time when we were meant to catch up on sleep, and we were just too shattered to do much anyway.

Chapter Four
BIG RIPPLES

....

And they weren't made by us

Though we were busy adjusting to the sudden onset of life as a family, the idea of making a film about the river never left us. It seemed that the more we tried to ignore it, the more it invaded our thoughts. Inspiration flowed past the front door every minute of every day. We often found ourselves talking about the kind of film we would make, what we would have in it, how we would shoot certain parts of the river and at what time of day the light might be best. Who would we feature? Which of the riverbank characters we were coming to know so well? We swung between enthusiasm one minute and despondency the next. Would we be taken seriously?

We had now witnessed the river's autumn and a snow-covered winter. We had seen ducks skating on the ice and kingfishers having their babies and fledging; we had seen otter and mink, grass snakes and voles; we had seen the river in torrential flood and on her twinkling best behaviour. It wasn't enough just to live beside her, we needed to film it all. By February we knew we could ignore the idea no longer and decided that if we didn't even try to get it off the ground we would never forgive ourselves. So we determined to get a shot, one shot of an otter, and hoped that would sell the programme to the people that ran the BBC's Natural History Unit. It was not as easy as it sounded.

Firstly, we had no idea how to predict when and if the otter would go by the house, or where else he might go. Otters have large territories and the likelihood was that ours would be using other rivers as well as this one. We had seen him on only a few occasions since that first time. Sometimes while staring out of the window into the darkness we had been alerted by large ripples on the surface of the water washing down the river ahead of him. Other times he would just arrive, popping out of the still water. But we were beginning to notice that this otter was a creature of routine and we had been doing some detective work.

Early on those frosty February mornings we would rush down to the muddy bank, often still in our pyjamas, before the dogs could get there and ruin what clues the otter might have left. We would steel ourselves for disappointment but sometimes, maybe once or twice a fortnight, depressed into the wet mud, would be footprints, typical, distinctive otter prints, rounder than a dog's and with five toes. Occasionally you could just make out claw marks on the end of each toe. The prints we were finding were large for an otter, almost three inches across, and that made us almost positive that the otter travelling up and down our river was the same large dog otter we had originally seen. Just to make sure, we used some plaster of Paris to take moulds of his pawprints whenever they appeared in the mud. We'd stand there shivering and mixing, shivering and mixing. I would hold the cut-off bottom of a plastic water bottle over the print to stop the bright white liquid plaster spreading across the surface of the brown mud and Charlie would pour it in. Then we'd stand there shivering and waiting until it was set. Soon we had quite a large collection of white pawprints cluttering up the spare room, and they were all the same.

As I read more and learned about otters and their movements I began to realise that their mysterious nature is half the fun and a large part of the challenge of studying them in England. In Scotland and Shetland life is much easier because they are clearly visible going about their business on the beaches during the day. Here however they are nocturnal, and the linear nature of their river home means they cover miles in a night, which was why we would rarely see ours. Even people who study otters all the time rarely see them and that is why they are obsessed by poo. It came as no surprise to me then when Charlie started bending down beside strange deposits on rocks by the river and sniffing them, although I did raise an eyebrow or two when he started bringing them home in jam jars and getting me to sniff them too.

If you see a man acting strangely on a riverbank be kind because it may be that he too is an otter freak and in lieu of seeing an otter will get very excited about seeing some poo. People in the know call it spraint, and the fresher it is the more excited they get. Unfortunately, although you can get an idea of how fresh spraint is by how moist it is, the only real way of telling is to have a good sniff. So, some mornings after a particularly satisfactory inhalation on the bank and a quick mix of plaster of Paris we would come in for breakfast beaming. It was wonderful to know that in the darkness the night before, while we had been sleeping, an otter had silently slipped by the house.

His tracks would tell us which way he had gone. When he was going upriver he would get out of the water below the weir, climb up the steps to the bridge, cross the front lawn and plop into the water behind the small reed bed opposite the living room. However, if he was going downriver we would find spraint as well as tracks and always in exactly the same spot. It varied in colour but never in smell. Once you have smelled otter spraint you will always know it again. It is fishy and dense, although many otter enthusiasts describe it as pleasant, and in the literature connoisseurs describe it as possessing the fragrance of jasmine tea. When I read that, I was briefly reminded of the ridiculous descriptions in the perfume catalogue on the flight that had got me into all this in the first place. To be honest, whatever poo you are sniffing, first thing in the morning it's all relative. Often you could clearly make out fish scales along with bones and other evidence of what our otter had had for his supper that night, usually trout or carp. We knew that otters loved eels, too, but we were not sure how many of those we had in our river.

To an otter, spraint is more than just a product of digestion; it marks his territory for the benefit of any other otters that might come wandering up his river. We knew our otter must have other regular spraint sites up and down the river – on rocks that stuck out of the water, logs or distinctive points on the bank, or even on ledges on bridges – and so for us the spraint was the most useful way of tracking him and trying to work out where he might be from day to day. We would be able to tell not only where he had been but when, according to how fresh it was.

It was when our otter was travelling downriver that we were more likely to see him. He would briefly surface under the bridge and then swim underwater past the kitchen window. At the top end of the weir he would clamber out onto the bank. This had once been a channel through which the water had run down to the mill but it had long been filled in and was now covered in trees and vegetation. Sometimes he would look around as we watched him; other times he seemed to be in more of a hurry, but he would always leave spraint before loping off into the undergrowth further down the river. Even if we hadn't caught sight of him, the next morning we would know that he had been past because of that spraint.

We decided on a shot that we thought could sum up the programme and impress the execs at the same time. We would try to get the otter on a trip downriver, as he paused to sprint on the bank just below the house. If we set the shot up right we would be able to get the house with the lights on in the background, which would show just how close otters were coming to civilisation as they returned to their previous territories, how much they had adapted and how, in some cases, they were living right under our noses. It was a safe bet because, as he paused to leave a sprint, we would get at least a couple of seconds of him out of the water. That way we would get a really good look at what we hoped would be the star of our show.

Night after night we waited. There was a small remote camera on a tree a few feet away from the sprint point, with leads stretching along the patio and into the house to a monitor. We sat inside and stared at that screen willing an otter to appear. The shot was lovely: the house in the background with the lights on, the river in the foreground and lit up, and the otter due to appear full frame. Every thirty minutes we had to remember to change the tape in the recorder.

As we waited for the otter to appear, we began to shape the programme and to think in detail about its structure and the story. During the day when Fred was sleeping we would cover the kitchen table with huge sheets of paper Sellotaped together, which we covered in turn with notes. There was no shortage of ideas. The huge paper tablecloth quickly began to reflect what was really happening – how Charlie, having loved this river since he was a young boy, had returned to it to find that the otter was also returning. By showing our otter passing through urban areas on the river, negotiating weirs and farms and quarries, we would show just how much they had had to adapt to make that comeback. Above all, we wanted to get across our passion for the river, which we would do by telling the story in the first person. We also wanted people to understand the thrill of seeing a rare wild animal and to feel it for themselves, so we both felt the film shouldn't be too slick. We brewed pot after pot of coffee on the hob as we pondered and discussed and concluded that sometimes as viewers we lose the thrill of the wild because it is too easy to access on TV; the glossy shots of creatures from all over the world fool us into assuming that it is easy to witness their behaviour at close quarters when really it is a tremendous privilege. We decided that with the otters we should focus on how tricky it was to find and then film them, so that if a shot was wobbly then the viewer would, we hoped, understand why and get a thrill from that.

Our film wouldn't just be about otters; we wanted to include the other residents of the river too. Being the organised one, I allocated each creature a different coloured pen and we noted the things about them that we thought were important or interesting. The paper grew as the ideas flowed: the fact that moorhen females fight for males, that ducks normally gang-rape each other but ours seemed to be a monogamous couple, the tremendous effort it must take a kingfisher to catch enough fish to bring up a brood of chicks. There were other aspects of the river that we were passionate about. We wanted to show its sheer beauty, the way it could captivate anyone who came close enough to stare. This could only be done by a very talented cameraman with a lot of film but I was assured that that would be no problem.

In the months we had lived beside the river, we had come to appreciate how remarkable each creature's survival was, given the constantly changing conditions. With each season and sometimes from day to day every river creature had to adapt to a home that was never the same. Holts and holes would be flooded one week and dry the next; the flow of the river might be impossible to fight one night and gentle the next. During mayfly season food might be abundant but in winter the river offered very little. We decided we wanted to get across the idea of a continually changing habitat and the effect this has not only on the animals themselves but on the way the whole system interacts.

After weeks of planning and reams of paper, we finally came up with a treatment – a condensed version of the programme which we envisaged. We printed it off on our best paper and sent it to various people at the BBC who commissioned wildlife programmes. We heard nothing but tried not to feel down; instead we concentrated on trying to get that shot. There was another reason that we kept looking very carefully at the prints and the spraints. For all we knew there might be more than one otter on the river. A few years earlier I had spent a day with the Environment Agency near my home town of Winchester in Hampshire, doing a crash course in otter tracking. Otters had recently returned to the area and spraint was regularly being found on the banks of the beautiful River Test, famous for its trout fishing.

The Environment Agency were having the spraint DNA-tested using a new technique being pioneered in Scotland. This was of interest to *Tomorrow's World* and they had sent me to film it. As we searched for spraint and sniffed various bits of bank, the very nice man from the Environment Agency, Steve, explained the importance of the DNA test.

'We know that we've got otters on the river but we don't know how many,' he said as he guided me to a stream off the main river. This led to a large, deep pond almost hidden behind a mass of vegetation and reeds. The mud around its edges was very thick and in the middle was an island, rounded off like a small hill.

'That's an artificial holt; that's where they sleep. Not many people know it's here. We built it for the otters and we know they are using it, but otters aren't very straightforward. They have holts all the way up and down the river and use different ones as they travel around their territories.'

'Are they all like this?' I asked, thinking that they shouldn't be too difficult to spot if they were.

'No way, this is a really des-res holt. They vary in sophistication from an old gas pipe, a hollow willow stump or a dug-out den with a chamber, to even a scrape in a reed bed.

Different holts are used at different times, depending on the degree of shelter they offer.' I was beginning to see some of the problems they were up against.

They had had some luck though. There is an old mill right in the centre of Winchester through which the river still flows. Later that day Steve took me down the steep, worn steps to the base of the mill. As the noise of one of Winchester's busiest main roads retreated, it was replaced by the loud whoosh of the river running through the mill-race below. The stone walls and floor were cold and slippery and the noise filled my ears. We crossed the river via a small wooden bridge which, although it had managed to stand for hundreds of years, felt very rickety, and got to a ledge at the far side of the building. There I could see a small video camera, the kind you would take on holiday, rigged up. It was pointing towards what was clearly a very popular spraint point on the end of the ledge. It was obvious that rather than face the racing water as it was forced through the bottom of the mill, the otters would hop out of the river onto this ledge, go along it for a bit and hop back in on the other side, just in time to rejoin the river as it swept under the busy road. While they were there they left their mark in the form of spraint. The Environment Agency had been filming them for some time with some success; there were certainly more than two otters brave enough to come through the middle of the city because one was very large and one was very small, but beyond that it was difficult to tell.

Next to the ledge was a window which now looked into the female changing room of the local women's YMCA. I wondered if anyone had ever looked up in the middle of changing to see a brown whiskery face peering in at them. Otters are, after all, renowned for their curiosity.

A few weeks after our search on the banks of the Test, all that spraint gathering proved worthwhile. The DNA analysis returned from Scotland with the wonderful news that there were many more otters than they had hoped for, possibly more than ten. So otters in Hampshire were doing very well.

Now here were Charlie and I, also trying to get the same kind of shot. All our hopes were pinned on it. The BBC might be unenthusiastic at the moment, but with that shot we could surely persuade the Natural History Unit to commission our film. We would have proof that an otter used our river.

There was another unexpected bonus. Film of an otter would also help inspire the flagging local otter group, who suspected that the otter was returning in increasing numbers but were working on a spraint-only basis. Charlie had been in touch with them the day after we first saw our otter, just to let them know for their records. Their first reaction was to be expected: polite cynicism. Because lots of people confuse otters with mink and phone such groups full of enthusiasm and buzzing with the excitement of their first sighting, they were slightly jaded but used to letting people down gently. So Charlie had explained that he was a wildlife cameraman and had seen and filmed very many otters all over the world; he knew one when he saw one. They jumped on him, in the nicest possible way. Before we knew it we were enjoying cheese and wine in one of the old college buildings in Bristol, and Charlie was entertaining them with an otter talk.

He was very nervous and tried to disguise it by moaning in the car the whole way there. Why on earth had he agreed to do it? Surely he could have nothing to say that would interest them. He was a cameraman; if he'd wanted to do public speaking then he would have been a presenter. Why didn't I do it? And so on and so on. My biggest concern was whether the group knew how long this might last; asking Charlie to talk about otters and not giving him a time limit could prove disastrous. I had no doubt that once he got started his nerves would disappear and the problem would be getting him to stop. I positioned myself at the back of the room so that I could show my support by nodding and smiling in the right places but was within easy reach of the door for access to toilets and, should he really go on, supplies of food and drink. In the event I needn't have worried.

The group members came from all walks of life, from teachers to salespeople, and were as keen as Charlie. He began by showing his slides from Scotland, including Shetland, and Peru, moved on to our collection of plaster footprints, and finally let them have a sniff of some spraint that he had collected in one of my jam jars, especially for them that morning. He even showed a slide of an otter delivering a spraint and a close-up of a fresh one. I wasn't sure they needed quite so much detail, but they couldn't have been happier.

At the end Charlie was bombarded with so many questions that he could barely get a vol-au-vent past his lips. The group were thrilled with the talk and even more thrilled that we had actually seen an otter. They had even laughed in all the right places. It might not be everyone's idea of a great night out but these people and people like them all over the country are the dedicated information-providers, crucial to the growing pool of knowledge about what is happening with the otter. They each have a river which they monitor, often daily, for spraint and tracks, and get no reward for their efforts other than the thought that they are contributing to the general knowledge fund. After our initial shock at their dedication, it came as quite a relief to know that there are others like us who sniff banks and forage around in riverside mud. By the end of the evening, we were determined to inspire them further with a shot of the animal in which they had invested so much time – proof that at least one otter was on their patch. We returned home invigorated and enthused.

So every night we continued to rig up the camera and all the leads, and every night we stayed up for as long as we could, keeping our eyes open with matchsticks and glued to the monitor. We even set alarms to make sure that we changed the tape in time. We had missed the changeover twice. The first time Charlie fell asleep and the second time I was ten minutes late. In that ten-minute window the otter might have been and gone. What could I do? I had the dinner on! I'd never make it in wildlife film-making if I was more worried about the spuds boiling over than getting the pictures. I could have cried. If it was this difficult to get just one shot, how would we ever make a whole film? Maybe we'd bitten off more than we could chew.

Meanwhile real life continued alongside the river. I read book after book about babies and cooked batch after batch of organic vegetable purée for our ravenous offspring. Green shoots appeared all over the garden and seemed to grow by the day, and then, as if by magic, great clumps of nodding white snowdrops sprang up all over the place. I discovered the joy of a new garden, the wonderful surprises that each season brings.

We got to know the M4 very well, and I was frequently up at five in the morning to whizz up to London for a *Tomorrow's World* script meeting or studio day. I was working on average four days a week, which seemed like a luxury to me. Before Fred, I had put in six or seven days a week for years, and although I love my job it was extremely refreshing to discover that there was life outside work. Charlie, a new baby and living in the countryside gave me a new sense of perspective; our riverside residence was working its magic on me and every day I was becoming more and more reluctant to leave.

We had no shot, no film commissioned and no sign of either, but life deep in the countryside by the side of the river was undeniably good.

In the big city life was good too. There were great celebrations at the BBC Natural History Unit in Bristol to welcome someone for whom both Charlie and I had previously worked. *The Natural World* series had got itself a new editor. His name was Mike Gunton, and on his new desk he had found our idea.

Chapter Five

REFLECTIONS OF
A KINGFISHER

....

Spring is launched in river tradition

It is no exaggeration to say that my first spring beside the river was a delight a day. This was my first spring anywhere in the countryside and I realised just what it really means. It is the start of the time of plenty, a signal that life rather than survival is about to begin, and, to most, that means breeding. In London I was only slightly aware of the change in seasons. It meant the difference between sitting outside or inside at the pub, the difference between taking a coat or not, a taxi or not. I was pleased to see the horse chestnut trees on Ealing Common blooming with bright new green leaves because I knew it meant hot days were on their way and the cherry blossom in spring on the tree outside my house would always cheer me up simply because it was so beautiful, but I never really noticed what the birds were doing. Beside the river you can't help but notice. It is like living beside the M40; you can't pretend it isn't there.

I was beginning to get to know the important dates in the river's calendar. In early March the kingfishers commenced their spring courtship displays with some spectacular high flying. I first noticed the change in their behaviour when I was bringing in the shopping, bag after bag, back and forth across the bridge – enough to feed a family of farmers, or Charlie. Normally, kingfishers fly quite close to the water's surface, often just a couple of feet above it. You can hear their high-pitched whistle and if you are quick enough you can catch the blue and orange reflected in the river. Then they are gone, negotiating the contours of their linear home as slick as a bobsleigh down a run, forcing you to be impressed, daring you to catch your breath.

My presence on the bridge that spring day was somehow getting in their way, and there seemed to be much more of a sense of urgency than usual. If the kingfishers are disturbed they take a detour out round the donkey paddock, wide over the driveway and into the next paddock before rejoining the river. This was happening today. I could hear the whistles but had to turn my head like an owl to see the dense sapphire body shooting like a bullet through the air, an almond shape with tiny wings, barely visible for their speed.

The kingfishers looked completely different flying across open ground and away from the silver strip to which they are so well adapted. The different perspective gave a much more powerful sense of how small they are – just a dot. It seemed miraculous that they could get from one side of the paddock to the other and remain airborne, they looked so frail. Guilt flooded through me as I realised that it was me, standing on the bridge gawping with my bulging carrier bags about to burst into the river, who was causing all this extra effort for the kingfisher.

In the house I kept watching as I unpacked the shopping. They were busy. Once they were back on their well-run air track above the river, dodging branches, speeding just above the spray, negotiating the tight turns, their size seemed perfect.

Any bigger and they would surely crash and burn. It didn't matter how many times I saw them shoot downriver through the tunnel of trees, their skill still made me smile as I tried to persuade the fridge door to shut.

So it was a real treat to watch them in their courtship displays. One particular morning they started just after dawn with a tremendous whistling which alerted me to the fact that something was up. I was out of bed in a trice. I was brought up on a council estate in Winchester and the importance of knowing your neighbour's business is in my bones. It makes very little difference to me whether those neighbours are animals or people – the slightest disturbance has me twitching my nets. So with no humans to watch I was the nosy neighbour peering out from behind the curtains, blinking a little at the bright day outside, which had most definitely begun.

Courtship displays can start as early as February, so we were a little late on the river this year. We also seemed to have more kingfishers to pair up. Even Charlie had commented that he was a little bewildered by the amount of activity we had seen the day before. This is an important time for all the birds using the river. If they want to reproduce, what they do now is critical; they must pair up and establish a breeding territory. It would be particularly difficult this spring because of the amount of competition; the previous year must have been a good breeding year. Charlie, who can distinguish one individual from another, had seen both males and females, familiar and strange, fishing from the bridge in the last few weeks.

Sometimes the way kingfishers flirt with one another is like young boys and girls in the playground, chasing each other up and down and then giving the one they particularly like a good bashing. But this morning I was in for a real treat, the high-flying style of courtship.

The clouds were high and brushed with colour, the light was golden, and all was reflected off the flawless mirror of the river, as pink clouds scudded above and below my window, the air coming in was sweet and the audience was settled and hushed. A glance behind me revealed the rest of said audience (Fred and Charlie) asleep and gently snoring in the big family bed.

The loudest whistling seemed to be coming from the eaves and I quickly realised that a male kingfisher sitting on the guttering a few feet above the bedroom window and full frontal to the rising sun. He was announcing his presence with all the joy that life could muster to the female on the bridge. (It's easy to tell the sexes apart, as females have an orange lower mandible, or lower beak, whereas males don't.)

So suddenly that I nearly ducked, the male left his perch and swooped, like some rebel Red Arrow, into his display. Up and down in dizzying dives over the pink and gold millpond, he missed its shining surface by a whisker each time. Then he landed next to his *amour*, and she took off at high speed. He followed, whistling, and I could hear their supersonic passage up the river, through the fresh green tunnel of trees. Then back again for her to perch and him to show off. She couldn't have been lit more beautifully; the orange of her chest was luminous, the blue on her back intense and her orange beak almost translucent as she bobbed in the warmth of the new day, turning her head at this angle and that as if to get a better view of her eccentric suitor.

He, meanwhile, was making enough racket to wake the neighbourhood, as he climbed again and again to the gutter and swooped off and over to the bridge as though he were on some invisible roller coaster. Nevertheless, oblivious and snoring, Fred and Charlie slept on.

Framed by curtains, peering through the window at the stage, I felt part of a drama. But this was real life and one that humans have no control over. It was important kingfisher business and I was lucky to see it.

They must have chased each other up and down the river at least five or six times before they settled on the bridge bobbing at each other. She moved sideways, her orange feet shuffling up the perch away from him. I smiled; we all know that game. But what did she really think? Would she accept him as a suitor? He couldn't whisper sweet nothings in her ear but he had another way. He focused on the river for around twenty seconds, her watching him, moving his head this way and that to get the angle which would allow him to see beneath the surface of the water. He dived.

It was one of those moments the memory runs in slow motion. Although over in an instant the vision is so stunning that it remains etched on the back of your eyes. He dropped into the water and then began to pull himself from its grasp with his precious trophy. For a second it seemed that he would never wrestle free from the weight of the river, but then suddenly he was out, droplets of water flying from his wings and body, the millpond a shimmering mass of star-shaped spangles. Holding it by the tail he thwacked his prize against the bridge, then, with careful movements, he adjusted the minnow in his beak so that it was straight and presented with the head first. He edged up the white handrail towards his orange-beaked beauty and with the river still sparkling and the dawn chorus playing in the background our hero offered his love an engagement gift. I held my breath.

She accepted.

With that formality over, the matter was settled. They were a couple, destined to spend the next few months doing the hardest work that any of us sign up to, and all in the name of reproduction. They would establish a territory, find a site and dig out a hole in the bank to nest in, lay and incubate as many as seven eggs and then fish, fish and fish again, to feed the hungry mouths. All this while hoping that no disaster befell them: no hungry mouths would find them, no pollution would endanger the fish supply and the lives of their tiny family and no other kingfisher would try to steal their patch. It was going to be a busy few months and they would need to bond well. Just to make sure he fed her a few more fish. Then they flew off. It didn't take much imagination to guess what they were doing...

The romantic opening to the day ended abruptly when I was called upon to change a nappy, but I still feel that I had witnessed a secret; that life on the river was starting a new cycle. It was strangely comforting.

Chapter Six
FLOODGATES OPEN

....

But no sign of Noah

By March we still hadn't got our commission, but we did know the new executive producer of *The Natural World* Mike Gunton had seen our idea. Because we had worked with him often before he had become quite a good friend and so we asked him and his family to lunch. I guess we were hoping, although this was never acknowledged, that if he witnessed for himself the life of the river he might be seduced into an instant decision. In truth our friendship probably made the project more problematic; you have to be very careful not to blur the lines between work and relaxation, and if you talk shop all the time his wife Sally gets cross.

Sally is tall and elegant with dark hair and big dark eyes. Like their three girls, Molly, Georgia and Ella, she is always animated and talking, although on our first meeting my overall impression was of sensitivity. She has the ability to work a situation out and sum up how everyone in it must be feeling, a knack of succinctly communicating her own thoughts and feelings on life at that particular juncture and an overriding sense of fun which suddenly bursts out from behind the thoughtful cloud with a big laugh. The girls are sometimes hard to separate; gangly and pretty they all adore Charlie.

In all the chaos we didn't have time to talk about the programme, and anyway we had invited Mike to lunch with his family not a business meeting. The house is fairly small and another family with three young girls filled it with noise and bustle, but Sally and Mike soon settled themselves in the kitchen looking out over the river. Fred thought his lucky day had come, grinned at all the girls in turn and then started again; we got busy with roasting trays, gravy boats and serving dishes and loaded the table with food. Big lunches and lots of people round the table had very quickly become a household tradition in the short time that we had been installed. If we had spent as much time decorating as we had done cooking and entertaining we would have been sitting in an immaculate kitchen. As it was, new cupboards had been squeezed in where they didn't quite fit and the paint was beginning to peel off the walls. Nothing really matched and very little was of our choice, but no one took any notice of these things; their eyes were always on the view.

Part of Mike's charm is his ability to make everyone feel relaxed. A smallish man with red hair (I would say ginger but I want to be alive after he has read this), he has a great sense of humour and endures lots of mickey-taking with a jovial smile. Before he got the job on *The Natural World* Mike had made a number of programmes about people and animals, including *Making Animal Babies* with me (me being the presenter, you understand) and one about Easter Island with David Attenborough. Mike cares about the people he works with as much as he cares about his career, which makes work all the more fun. He is also very creative and has a great understanding of what people want to watch on television. However he is completely disorganised; he has a diary but only ever really looks in it to find out where he was meant to be half an hour ago. When he doesn't return an e-mail it isn't because he doesn't want to but because he hasn't read yours yet. If he says he will meet you at a given time and place you are responsible for getting him there by means of several phone calls to his fantastically efficient assistant Barbara, and even then you must accept that he will generally be half an hour to forty minutes late, full of apologies but late nonetheless.

As we all squeezed round the table Charlie and I were finding it difficult to hide how excited we were about the possibility of filming on the river. We were also wondering whether to tell him that we had found a second set of otter footprints — a smaller set. But would that sound too much like work? We had only discovered these in the last few days and at first we had not been completely sure, but this morning's prints had been very clear. We had two theories. The prints could belong to a smaller dog otter encroaching on our large dog otter's territory, but it was unlikely that he would tolerate that; a male will only accept another male on his territory if it is a cub. So our second, more exciting theory was more likely — that it was a new female. Male and female territories overlap, and although otters rarely encounter each other in the flesh, they will put up with sharing fishing rights.

But before we could steer the conversation round to otters, we had a surprise visit from another rare couple. We were just finishing the main course when a pair of goosanders flew down onto the river.

To me, this event was no more than the appearance of a couple of exotic-looking ducks, with nice pink breasts, that we hadn't seen before, so I was preparing a 'How nice, dear' kind of comment, a 'Look at the two ducks, children' kind of moment. But before I could open my mouth Charlie had leapt from his chair and, remembering just in time that Mike's three girls were sitting around the table before he swore in his excitement, told us that these birds were usually only found on big lakes and the coast. It was very rare to see them on a river, *very* rare, he repeated, in a state of heightened animation.

Our eyes met across the table. Was this an omen? Surely Mike would be so impressed by our wildlife oasis that he would commission the film straight away. But the organic beef was just too good, and with a couple of nods and smiles he acknowledged the presence of the dancing goosanders and went back to wrestling with one of his girls for the last Yorkshire pudding.

Late that afternoon, after lots of joviality and promises to return from the Guntons and tears from Fred, we waved off our visitors. The house was very quiet. We had managed to forget about work for a few hours but suddenly we remembered. Depression set in. Then we felt bad because it was only meant to be a friendly lunch anyway. It was the weekend; what did we expect, a commission? So we went for a walk in the rain.

The river is extremely good at exaggerating the mood of the moment. If it is a gorgeous day the river reflects the blue sky and sunlight and adds another dimension of sparkle that makes you glad to be alive. However, a miserable day can be made worse by a walk beside the river. The drops descend on the green-grey water and then ripple away to nothing, the river and the bank are one big mud bath and all the birds that normally dance up and down singing and amusing us seem to disappear. At times like these you want to believe in those wonderful children's books which depict cosy tree-trunk homes with mother birds dressed in aprons carrying wicker shopping baskets and making everyone cocoa. You want to think that all the creatures have somewhere warm and cosy to retreat to rather than just an old gas pipe or an unsheltered nest.

Cold and wet, muddy and slippery, with no sign of an otter, the walk was doing very little to lift our spirits and the rain showed no signs of letting up, so we decided to go home.

Yet it was only because of the rain that Charlie went outside later that evening.

Once, when we had only just moved into the house, the river nearly flooded. It can be pretty scary; the river becomes a raging torrent and could sweep you over the weir and dash you onto the rocks below in seconds. Normally it flows a few feet beneath the bridge outside the house, but when it fills up in the flood season the water is only inches below your feet as you cross. Charlie had got very nervous and phoned the Environment Agency to enquire about sandbags. When the tired and overstretched woman on the end of the phone asked how close we were to the river, he just opened the front door. The only thing she could hear was the roar of the water and it was all Charlie could do to persuade her not to call the emergency services.

It had been raining all day, and the river was already high, so that night Charlie had to raise the sluice gates just outside the house so that it wouldn't burst its banks. Suddenly he heard an unmistakable, high-pitched squeak, almost a whistle. It was an otter.

Standing downriver, Charlie shone the torch back up towards the house. What he saw, caught in the beam, was almost unbelievable: a female otter with two cubs, just at the end of the weir race outside our kitchen. It was the same spot where the male used to spraint. The cubs were tiny, barely as big as my hand, probably not really big enough to be outside the den, and the mother looked very nervous. As soon as she saw Charlie, she grabbed the cubs and retreated into the woods on the riverbank.

Charlie shot back into the house breathing so hard he could barely tell me what he had seen. We switched off the house lights but left the river lights on and, crouching down by the kitchen window, we waited to see what the otter would do. We could hear the babies whistling the whole time. My heart thumped. Otter cubs on our river! Maybe we would even get the chance to film them.

Suddenly, out of the shadows, there she was! She had made her way downstream through the trees and then crossed the river out of sight of the house. Now we could pick out her dark form making its way along the opposite bank, the bank below the weir, which is covered in trees and vegetation. She wasn't moving very quickly and we soon realised why. She was carrying one of the cubs. It swung from her mouth like a tiny kitten, and as I watched I could almost feel the anxiety radiating from her. She was putting both cubs at risk. As she crossed the small stream that runs into the river opposite the house, we realised that she must have left the other cub somewhere. She climbed the steps leading up to the bridge and the footpath and disappeared into the shadows.

We could hear it straight away – a persistent, insistent whistle, the calling of the cub she had left behind. I wanted to rush out onto the patio and tell it to be quiet. If we could hear it, any fox or other predator in the vicinity might hear it too. The water level on the river was rising. We could see the difference from just ten minutes before and that meant that the flow would be stronger and much harder to swim through, particularly if you are carrying a cub.

We knew that the mother must be trying to find a safe place for the first cub so that she could come and retrieve the second one, but did she have somewhere in mind or was she desperately searching right now? If she had been on the river some time she would probably have other holts or lie-ups further upstream, but if she was searching for somewhere new then she might take a long time. The minutes were ticking by and the whistling continued, loud as ever. What if she didn't come back? Visions of a hungry otter cub swimming around in our bath battled with those of a hungry fox picking up the noises of the vulnerable baby. Our kitchen was silent and dark; we didn't really know what to say or do. All we could hear inside were Fred's hiccups as he happily watched the water flow by the window; all we could hear outside were the roar of the river and the whistling of the cub.

Charlie decided to go outside to see what was happening to the river level. I stayed in the kitchen and then thought of something useful. When you are the other half of a wildlife cameraman you feel a bit foolish with a camera in your hand, but I suddenly remembered the little video camera that we had bought for filming Fred. There isn't enough light, I said to myself. You can't film an otter at night with that camera – the quality won't be good enough. You'll never see her through the viewfinder. She might not even come back … That final thought made the decision for me. It was far better to be doing something than staring out of the window at the rising water, listening to the poor cub and trying to imagine what on earth the mother was doing. So I grabbed the camera out of the kitchen cupboard and ferreted around for a charged battery, then a tape. Then Charlie came back in. I began to explain that I knew this wouldn't work but he interrupted: 'Brilliant! What a good idea.'

I shut up and handed him the camera.

We sat for what seemed another age listening to the roaring and whistling, Charlie with the camera on 'record' pointing at a spot on the bank where he suspected the mother otter would emerge. I was sitting in the dog bed again, and Bill, evicted, stared at me from under the kitchen table. I don't think he minded but he probably wondered why I was using his bed when mine was so much nicer. The squeaking seemed to be getting louder.

All at once she was there, not even a ripple to warn us. Her slim, brown form slipped out of the water at exactly the point Charlie had predicted, then she was gone. One, two, three seconds later, the whistling stopped. Mother and cub were reunited, and tears of relief pricked my eyes. But then I realised that if one cub was with his mother then the other was alone. We desperately scanned the opposite bank. A fleeting shadow told us that she was already on her way to reunite her family. The water was rising but we were sure she had found a safe holt.

For now all we could do was rewind the tape and skip around the kitchen celebrating our first shot of an otter. We had it, at last, after all those months. It lasted approximately a second and a half, was fairly dark and you couldn't make out much detail, but you could clearly see that it was an otter. She looked frightened and frail and urgency filled her every movement. But we had it, and we loved her.

We immediately phoned Mike, who was either nursing a hangover from our boozy lunch or already in bed. Either way he sounded dozy.

'What? Tell me again, but slow down.'

The excitement made Charlie's voice tremble as he told the story again.

'I can't believe it, just where we were sitting today, and you've got the shot. That's fantastic! I'd love to see it. You must be so chuffed.'

But then we heard nothing. Eventually we gave in and phoned Mike's office, asking him what he thought. He said he'd let us know by Easter. Definitely.

Chapter Seven
HALCYON DAYS

....

Now there was no looking back

April came, we still hadn't got a commission, but we did have a moorhen. She was called Mary and she nested in the reed bed opposite the house. We could watch her from the sofa as she strutted around on gawky green legs and it was too easy to imagine that she had been poking her beak into everyone else's business and was now off to gossip. From the flutter of her wings to the twitching of her head the vibe she gave off was busybody.

Every morning through late spring she tried to raid the kingfisher feeding pool. This is what we use to bribe the kingfishers into feeding close to the house. It is a small pool, about the size of a hanging basket, suspended in the river beside the bridge with the help of some buoyancy aids, and is always well stocked with fish, the maintenance of which can be a full-time job in the spring and summer. The first can jump out but they rarely jump in. The idea of the pool is to encourage the kingfisher to sit on the bridge railings and fish. If it is feeding a nest full of young, even a small bird like the kingfisher needs to catch about sixty fish a day. Occasionally we put it out during a flood or in the middle of winter, when it can be a lifesaver. If the water is full of silt and mud or flowing too quickly the kingfisher can't see to fish in the deeper areas and becomes a frequent visitor.

Charlie had invested a lot of time and effort – while sitting in the sunshine on the patio – in the design and structure of the fishing pool. It had to float at just the right depth for the fish to be able to get in and out but be shallow enough to entice the kingfisher. So he found it intensely irritating when our new neighbour, having watched his antics with interest, took advantage of his ingenuity. Mary, it seemed, was exceptional in the truest sense of the word. She always had been a little different. One day we even saw her up a tree. Since when do moorhens perch in trees? We could only conclude that she was being especially nosy.

Every morning that spring there was a race between human and moorhen to determine whether or not Mary got fish for breakfast. Just as the sun cleared the horizon and it was starting to get light, just as a gentle mist began to leave the surface of the water, a dark shape could be seen lurking in the shadows of the reed bed. Sometimes even the odd chirrup could be heard. When she had checked that the coast was clear, Mary slowly and deliberately stepped into the cool water. As this happened there was a snuffle and a snorting in the bedroom, the strange lump underneath the duvet started to move, and at the sound of Mary's chirrup it was as though an electric current had been passed through the bed.

'That bloody moorhen!' went the battle cry.

Two long skinny legs thrashed around until they found the floor and then the chase was on.

At this point, smiling in anticipation, I moved to the window for the morning show. Mary swam towards the fish pool, glancing left and right in a surreptitious manner.

I could hear the progress of my other half through the house. Bang, thud, bang – the stairs taken in two leaps.

Mary's long green legs were pulling her over the lip of the pool.

Crash, tinkle, tinkle, curse – the hall door flung open and the front door key falling to the floor.

Mary, like a little old lady in an old-fashioned black swimsuit at the swimming baths, had just settled herself into the pool.

The front door was wrenched open and there on the patio below me was a naked, crazed man, a cross between Victor Meldrew and Basil Fawlty with bed hair.

'Get off those fish!' Charlie yelled, waving his long arms wildly as he ran towards the river.

Mary always looked deeply shocked, mortally offended and disgusted that she should be spoken to in such a manner, and she fled, running across the surface of the water screaming rape and pillage until she reached the safety of the reed bed.

Charlie always looked a little shocked when he came to himself, as if he had been betrayed by his baser instincts. He rubbed his head, turned round and went to put the kettle on. Watched all the while by a beady eye from the reed bed.

That first spring passed quickly; there were new things to see every day and daffodils in every corner, the river flooded a couple of times in the spring showers but not as dramatically as it would in the autumn. The birds were frantically busy. Spring is their season and they were everywhere you looked: flirting, feeding, mating. The ducks were always hungry, begging for food outside the kitchen. One duck had had a huge brood and her ducklings would follow her under the bridge in search of breakfast, one yellow bundle of fluff after another until the last one, who was black. He was the only black duckling on the river and always last because he was so inquisitive. Being distinctive earned him a name; he was called Trevor.

The moorhen stepped up her campaign, swimming up and down the river and trying to raid the fish pool every five minutes. Blue tits nested just above the front door; coal tits and long-tailed tits called to each other. By 8 April we had even seen our first pair of swallows. Things were moving quickly. It seemed like every five minutes that the grey wagtail trilled from the bridge, loudly proclaiming his territory to anyone who had the time to listen. And then, with his beautiful mate, like a pair of fairies, they would flit over the weir and up and down the river, flirting and mating the whole morning. When the romancing was done you could see them with tiny sticks or mouthfuls of dog hair stolen from the patio stuffed in their beaks, busier still, building a nest in the ivy beside the sluice gate. It was just out of reach of the spray, and from the house we watched them going backwards and forwards in and out. If you leaned over the sluice very slowly and carefully when you were crossing the bridge and looked down, you could just catch a glimpse of the female testing the nest for size.

By the time spring was turning into summer Charlie and I had reluctantly given up on hearing anything back from the BBC about our project. Even Mike had gone quiet on us. We decided to try just to enjoy the river without constantly thinking how we could capture it on film and what a fantastic programme we would make. We thought perhaps that it was just our own love affair with the river which was making us see it through rose-coloured spectacles, that maybe other people could never really see it as we did. What did we know about what made a good film anyway? We agreed we wouldn't ask Mike about it any more. Maybe he was just too embarrassed to tell us that it was a rubbish idea. After all, as a friend of ours, he was in a very difficult position.

As so often happens, as soon as you let something go it comes back. No sooner had we made that decision than we had a phone call from Mike's office to ask if we would like to come in for a meeting. What did this mean? Did he like the idea after all? When, a few days later, we walked up the corridor to his office we were very nervous. We had both worked with Mike so often that we are very comfortable with him; it was just the situation which made us feel awkward. We had no idea what we were doing there or what if anything we should have prepared. So when we began both Charlie and I over-compensated by talking too much and cracking stupid jokes, most of which involved insulting Mike.

This joviality was cut short quite quickly and Mike began by talking about budgets. Neither of us knew the first thing about financing a film or costs – there were some post-production activities listed on Mike's budget sheet that we had never even heard of. We had a lot to learn and were up front about what we knew and what we didn't. We admitted that we could only make the film if we had lots of support. That, I think, rather than going against us, by some curious twist seemed to make up his mind. That, and the fact that we agreed to make the film for an absurdly cheap figure.

The bottom line was that I would barely be paid at all for my contribution and Charlie would be on less than half his normal daily fee, certainly less than the assistant cameraman would get. Between the three of us we slashed the budget: no transport costs, no accommodation costs, no publicity costs thanks to a friend of mine; we would get a great rate on equipment hire thanks to a friend of Charlie's. The only thing we didn't skimp on was money for special equipment to get memorable and different shots. But we could hardly think – our heads were spinning with the excitement. We didn't care about the money – we were being given the chance to make our dream film. We were television producers! Or at least we had the potential to be. By the time we left that office, we didn't even care about the Bristol rush hour.

Looking back, I wonder how Mike must have felt about giving two complete novices the chance to make a television film. I know him well enough to believe that he must have been confident we wouldn't waste the licence-payers' money, but he must have had some doubts.

We floated home, exhausting our phone batteries by ringing everyone we knew with our fantastic news. Back at the house we did a lot of dancing around the kitchen with Fred and opened some champagne. I remember that our hands were shaking as we clinked our glasses together. The next day was different. We woke up still floating but slowly it began. Life began to feel bumpy again. Our heads began to feel fuddled and it wasn't the champagne from the night before.

I found a wet towel on the floor in the bedroom.

'Who left this here?' I asked slightly irritably.

'What?'

'This towel.'

'Who do you think?'

'Well, it wasn't me and it wasn't Fred.'

'Well, why are you asking then?'

'Was it you?'

'Of course it was me. It wasn't bloody Bill, was it? He doesn't go upstairs to wash his arse and use a clean white towel and then dump it in the bedroom, does he? Give it to me.'

'No, I'll do it.' The sigh may have been unnecessary.

'Why are you being a martyr? Give it to me. I'll put it back.' He grabbed the towel, marched into the bathroom and shoved it roughly onto the rail. I chose to ignore the fact that it wasn't folded properly and decided to do it later. When I get stressed I can't cope with the slightest bit of mess.

'What?' he said, looking at me.

'Nothing.'

Now it was his turn to sigh. 'I suppose it's not folded properly.'

'It's fine,' replied the martyr.

Taking the towel, he began to fold with grand, dramatic gestures, then replaced it on the rail. 'Is that better?'

'Yes, thank you. Although I don't know why you can't just

do that in the first place and save a row and me a job.' The tone was getting patronising. I could hear it, I just couldn't stop it.

'Well, I don't know why you can't just ignore it.' The volume was beginning to rise.

'Because they smell if you leave them.'

And we were off, down a familiar route, and neither of us could let it lie. Eventually we resorted to insults and shouting.

'Perhaps if you cared about me a little bit more then you would see things instead of tripping over them, or are you just leaving them for me to clear up?'

'Oh, and now I suppose you're going to bring up the time last year when I left the clean washing outside in the basket all night and it rained. You're so petty!'

'Well, you just don't care.'

'I do care. I just don't always show it. I'm a bloke.'

'Oh, like that's an excuse for everything.'

We argued about everything and anything we could think of, but not the programme. We didn't even mention the programme.

Every couple argues and so they should. Show me a couple who don't argue and I won't believe you. Our rows tend to follow the same pattern. I accuse Charlie of not doing enough housework, or of leaving his stuff lying around and he accuses me of not being perfect either. I begin to rant as I storm around the house finding more and more things in the wrong place (some of which I have put there) and he begins to sulk because I am ranting. Then we both sulk for a bit until one of us gives in. Often I cry at this point.

A few hours later, when we were speaking again we both confessed to being terrified. It's all very well being given the break to make your dreams come true, but then it's all up to you. The project was actually happening, reality was kicking in and the idea of failing was unbearable. But we had managed to persuade Mike to commission a film that we had no chance of making. Our minds were numb with panic. It felt as though we knew nothing about television, not the first thing about making a film. We had to deliver a fifty-minute film to the BBC by October 2002 and neither of us had a clue where to start. I had really only presented – you could hardly count me as an experienced director – and I had never produced a thing. Charlie was a great cameraman but he had never produced a thing either. To cap it all, it was May already. Spring was almost over. As usual, Mike was late; we had missed it!

And yet, although we were too busy panicking at the time to realise it, this is when everything actually fell into place. We thought we were starting from scratch, we thought we didn't even know where to start, but now I can see that we had already started. We had just spent almost a year researching – we knew the major locations well and where to find our main characters, we had talked endlessly about the style of the programme and about different shots and subjects. And we had the rare advantage of not having to plan a filming trip, pack lots of gear and work out the logistics before we started filming, which is what we would have had to do if we were filming anywhere else. Our location was our own doorstep; we were ready to film as soon as we walked out of the front door.

After a while the panic began to subside. Charlie began to breathe again and I was preparing to present Crufts for the BBC. Although the dog show is usually held in March, it had been delayed because of the foot and mouth epidemic.

Spring was rapidly becoming summer. Mistle thrushes were nesting in the tree opposite the kitchen and a squirrel was constantly leaping over the river from tree to tree, carrying nuts and bedding. We had begun to plant up the garden. I made a herb garden and put in lots of rambling roses and we pulled up a thirty-foot conifer that had overstayed its welcome in the front garden. This was pretty exciting because our neighbour Pete, who is a builder, arranged for his mate Steve, another neighbour, who hires out heavy machines, to send his mini-digger up the lane to get the root system out, leaving us with what Pete terms in his broadest Bristolian a ruddy great 'ol'.

Fred thought this was one of the best things he had ever seen, judging by the way he waved his arms about.

All this time marching meant that we now had only one spring, that of the following year, to shoot everything that we needed to include in the programme. There would be a lot to cover in just a few weeks. Everything would be riding on that one short season and if the weather was bad we would have to write it off. However, there was nothing we could do about it now so we decided that Charlie should start filming kingfishers as soon as possible to make the most of the tail-end of this spring. We had missed filming the courtship displays this year, but if we were lucky we would catch up with our nearest pair in time to film them nest-digging and having chicks.

We also needed to catch up with the otters; we hadn't seen them in over a month. We lived with the worry nagging at the back of our minds that perhaps mother and babies had not survived the flood. We had, however, seen quite a bit of the mink.

The mink has a terrible reputation, and most people sneer at the very mention of its name. The European mink is now rare, but the American mink has become well established in the UK partly because of fur farms which, although now illegal, used to breed the American version for their pelts. The American mink was the most common fur-farm animal, and in the 1980s almost thirty million pelts a year were produced worldwide. Many mink escaped from farms, still more were 'rescued' by animal-welfare campaigners who, from the 1920s, set them free to begin a new life on British rivers. Although this liberation was well-intentioned and great for the mink concerned, it was a mistake. Mink have survived well in Britain but they are not a native species and the river ecosystem has not evolved to take them into account.

They are hated primarily because they are such efficient killers. Mink will take chickens, ducks, ducklings, fish, almost anything they can find. There is also evidence to suggest that the decline in the water vole population and in those of many native birds is due to predation by mink. Yet I don't think that you can blame the mink itself, an illegal immigrant trying to earn a living after fleeing from a fur farm. Now, the mink is just the same as all the other creatures on the river, and, dare I say it, when you look at them, when you see beyond all the prejudice, they are a little bit cute.

An American mink is not nearly as big as an otter, which is usually twice its size and can weigh ten times as much. And of course it is not nearly as attractive. But if you have never seen either in the flesh it is easy to mistake a mink for an otter and the two are often confused. As well as being smaller in body the mink's facial features are more petite and more pointy, and it has a much smaller tail. Whereas the otter's tail is thick, long and tapering, the mink's is fluffy and cylindrical. The mink is also generally much darker than the otter, although one we had been spotting lately had been almost grey.

We wondered whether it was significant that we were seeing the mink in the otter's absence. As the otter returns to our rivers there is much talk about what will happen to the mink. Will the two creatures compete for the same food? Will they fight over territories and if so who will win?

Early research seems to indicate that direct competition for fish will not really be an issue as mink are more dependent on other terrestrial prey. They are not nearly as well adapted to fishing as otters and would probably lose out in the fishing stakes. As for territory, we had witnessed on our river as other people had seen elsewhere that the territories of the mink and otter could overlap. We even suspected that the mink might use otters' holts when they were not resident themselves, although we had no proof of this yet. There were also rumours of otters being a physical threat to mink. One story had come from an otter watcher in the north of England, who claimed that he had seen an otter eating a dead mink, and there were reports from Russia of mink remains in otter spraint. It would be interesting to observe what would happen to the much-maligned mink if the repopulation of the otter proved successful.

At the moment the river seemed to be particularly busy with mink. There were at least two, the almost grey one, and a smaller dark brown one, with a more weasly-shaped face. They were very bold. On a sunny day in April Charlie had filmed one of them on our home camera. It had been making its way up the river in the middle of the day without making any attempt to conceal itself. It came up onto our patio, realised there was nothing of interest there and wandered across the bridge. We lost sight of it for a while as it clambered into the ivy growing around the sluice gate but it soon plopped into the water on the other side. It nonchalantly made its way across the pool and skirted the reed bed. It didn't have the faintest clue that we were there, and not a care in the world.

Charlie squeaked it, which is something that wildlife cameramen often do to get the attention of a predator like a mink. He sticks two fingers in his mouth and, instead of blowing, draws breath in and makes a loud squeaking noise. It is really piercing and can carry for quite a long way and still be heard despite the noise of the river. The theory is that it mimics the sound of an injured rabbit or some other small creature and tempts the curious predator back into range on the promise of an easy dinner.

Our mink immediately dived into the shelter of the reed bed. We wanted to try to get ahead of him to film him coming up the river and so we ran quietly into the paddock upstream. There we nestled down in the long grass on the bank and waited, camera running. There was no sign of him – perhaps he had become suspicious – so Charlie squeaked him again. Suddenly we could see the arrowhead of ripples which showed that our mink was now making his way towards us. The squeaking seemed to be working, but he was still cautious, hugging the bank so we couldn't see him. Charlie continued to squeak but then the ripples vanished. The mink must have got out of the water and was on the bank. We listened for the sound of him coming through the grass. There it was, a little rustling not far to the left of us. We kept as still as garden statues. There it was again, but higher up somehow. Charlie squeaked again and above us there was a rustle, a face and then a crash and a splash. The mink had tried to climb up the willow tree next to us to get a better look at these huge creatures who sounded like rabbits. However, he was so busy being nosy that he had missed his footing and fallen with an enormous splash into the river. His cover was blown and so was ours, and while we laughed in the long grass, he swam away as fast as he could up the middle of the river. I know that we shouldn't impose human emotions on animals but I am convinced that was one embarrassed mink!

So we had plenty of mink but no otters, spring was almost over and we had a film to make. We wanted this film to be full of spring action yet had hardly shot any, and it was now time for me to head off to Birmingham to film a load of baying hounds and their owners. Meanwhile Charlie was left holding the baby. The weather was perfect, just right for filming, clear blue skies and plenty of light, but it was very difficult to hold a camera steady with a baby, however delightful, in the other hand, a baby who by now was spending every waking minute trying to crawl. Fred had already spent a fair amount of his life in the Natural History Unit stores, where Charlie would pop him on a bench and do deals while changing his nappy. Neither Nick nor Luke, who ran the stores, were as yet fathers, and were so distracted by the contents of Fred's nappy that they would stand there with their jaws open and unwittingly agree to hire us lots of expensive equipment for very little.

Charlie had taken fatherhood pretty much in his stride, and so far had been able to sling Fred on his back in the baby carrier and take him filming. Now Fred was having his own ideas about what he wanted to do and they didn't always involve going in the same direction as Dadda. This week, though, Fred and Dadda would have to sort out their differences between themselves.

Crufts is a funny event. My first view of it is the same every year. My hotel room overlooks the lake and the front entrance to the NEC, and on the first morning, which is always a Thursday, I open my window and laugh. Along the path, in their hundreds, dogs of all shapes and sizes are arriving. Sometimes they are being pushed in huge trolleys, sometimes they are allowed to walk while their owners push the trolleys beside them. The trolleys contain all the essential grooming equipment: hairdryers, different types of brushes, sprays, scoops – you name it, it's in there. There are more tarting accessories in each trolley than in the whole of our make-up caravan. You see big, fluffy, bearded collies, their coats streaming back behind them as they parade like incredible canine shampoo ads, with socks on so that they don't get their feet messy. Terriers yap from the inside of their travelling cages as they are pushed past the lake. Gun dogs stride alongside owners decked out in plus fours. And yet, however beautifully trimmed, groomed and trained, they are all dogs, and even the purest white poodles sniff as many bums as they can on the way in.

For those dogs showing early, there are practical matters to attend to. This is their last chance to relieve themselves before the big moment and owners hang around with their charges on leads in one hand and a carrier bag in the other, despondent looks on their faces, hopefully visiting sites that might inspire a 'movement'. Those lucky enough to see some success then proudly transport the doings at arm's length to the special bins before taking a deep breath and entering the halls.

I love presenting Crufts, but it does have its critics. Nicky Campbell once asked me why I was interested in such a 'fascist' activity and yes, I'm sure there are bad breeders who care little for dogs and more for money, but there are bad people in every field. For me, what is delightfully eccentric in people is brought out at Crufts. And then there is the love of animals. It is expressed differently from the way most people express it but everywhere you look you see genuine devotion to dogs.

My favourite part is watching the Border collies doing the activities. Knowing collies as well as I do, I appreciate how much joy they get out of the agility and flyball games. They bark all the way around sometimes and their tails wag so much they nearly fly off, yet they are still focused, still eager to please and so full of fun. You can't help smiling as you watch them whizz around the arena. This is a dog that was bred to work, except they don't know the difference between work and play. So, they couldn't have a nicer time, and now that the message has got through that dogs perform better when trained with positive reinforcement – that is to say praised for doing well rather than shouted at for doing something wrong – there is never an excuse for an unhappy dog or an unhappy owner.

Crufts is a feast for the eyes. I often wish that I had the time and the freedom just to wander around and take lots of arty black and white photos – so many owners look like their dogs, which must mean I look like a Border collie. Some of the pampered pooches are snoring away while they have their hair done and others are getting a last-minute cuddle. Here a woman leans against her Great Dane, passing the time between rounds by reading a book, there a child is asleep in a dog bed. Look one way and someone is winning their first rosette, look the other and a Yorkshire terrier is attempting a stand-off with a German shepherd police dog. There is always something to see, something to make you smile.

Nearly every day is busy for the BBC team and that is why my memories tend to be snapshots, images picked up as we make our way through the crowds from location to location.

The team is fantastic. All the BBC events people know what they are doing as it's got to be one of the toughest jobs in TV turning up with a massive crew and covering an event live, giving the audience at home a real sense of what it is like to be there, making them feel that they *are* there. This means shooting and editing films in no time at all. Sometimes big chunks of the competition are still being edited when we are already on air and no one knows exactly how they will turn out. That is when the commentators come into their own, with their ability to talk with ease about any dog that appears on the TV screen in front of them. There are 137 breeds to know about, and over just four days 22,000 show entries will pass through the doors of the NEC – that's not including the 2,000 dogs in the other classes: obedience, flyball, agility and so on. Each day passes rapidly as we gather information and film for the evening's programme, write scripts, help each other out, prepare for the moment that we go live into people's living rooms and let them know how it was at Crufts that day. Whether they are passionate about dogs or really couldn't care less, we still want viewers to enjoy the programme and be entertained by it, and finding the right balance is one of the hardest tasks.

So I usually keep my mobile switched off as I have so much spinning around in my head that I can't cope with the distraction. However, this one day I happened to have it with me while I was filming. It was in my pocket and, even more sinful, it was switched on. We weren't rolling when it rang, so I answered it. Charlie, breathless, was at the other end.

'I'm getting the most amazing shots of my life.'

'What?'

'Two kingfishers trying to drown each other. Got to go.'

And with that the line went dead. I didn't even have time to ask where our son was while he was getting these shots or to get much idea what was actually going on. I knew better than to call back.

Ten minutes later the phone rang again. This time I could

hardly understand what Charlie was saying because I was standing in the middle of a horde of overexcited shih-tzus and their owners. First there was just heavy breathing and when he did finally manage to speak his voice was so trembly it was difficult to pick out the words.

'I really have just got the best shot of my entire life,' I managed to make out.

I smiled. 'What, two kingfishers trying to drown each other?'

'No, even better.'

What could be better? This is where the dropping adrenaline levels kicked in and made him sound like he was talking rubbish. Either that or I was going out of range, so I stopped him. Around me the bemused crew were waiting to start a piece to camera. I said I would call him back.

I waited until I was on a break, back in the peace and safety and full service of the make-up Portakabin. It seemed like an age but it was only twenty minutes later. Charlie still hadn't calmed down.

'Right, tell me what happened.'

He had been washing up at the kitchen sink with Fred playing at his feet when he had become aware of a lot of kingfisher noise outside. Most people would barely have noticed it, but after so many years working on the river Charlie's ears are so attuned to kingfishers that he notices them even if he is indoors with Bob Dylan on full volume. When he looked downriver out of the window, he realised that a fight was brewing. It seemed to be centred around the sandbank where kingfishers nest most years. There was lots of high-pitched whistling and flying to and fro. With fingers crossed, he dashed next door with Fred.

As ever, Delia and Fred were thrilled to see each other.

'I can't think of anything I'd rather do than have a little visitor for half an hour. We have composing to do on my electric organ, and Beethoven to listen to,' she said as Fred was flung into her arms.

The kingfishers were so absorbed in their fight that they didn't notice Charlie and his camera and tripod sneaking into position on the opposite side of the river just below the weir. Or if they did, they didn't care. It was two females, one from our couple and an intruder. Our couple had already started nest digging and had chosen this site, close to the house, just downstream from the weir.

This is an ideal nest site, which is why it has been used by kingfishers for generations. If you see that the water has carved a high muddy or sandy bank beside a river then it is likely that it is being used by kingfishers as a nesting place. Look for small holes, and the one with lots of droppings just outside it is the one that has been used most recently. A glance at the height of the hole will show you that the spot has been carefully chosen to be high enough not to flood but still safe from predators – mink have been known to raid kingfishers' nests.

It can take a pair two weeks to dig a nest. They take turns burrowing into the mud with their beaks until they have a long tunnel with a chamber at the end where the female lays a clutch of up to seventeen eggs. Sometimes they invest a lot of time digging a tunnel only to hit rock and find that they can go no further, and then they have to start all over again. They are only tiny birds and it is a remarkable feat to dig such nests.

When the dispute arose, our pair couldn't have been digging for more than a day because they hadn't got too far into the bank. We had seen a lot of kingfisher activity on the river this year and it seemed that this was causing problems: too many kingfishers and good fishing territory becomes hard to find. The intruder on the patch had seen the whole package – the territory, the male and the nest site – and decided to challenge our female for it. At first the home pair merely tried to chase her away in between digging, whistling loudly in time-honoured kingfisher tradition to warn her what would happen should she persist, but the new female must have been desperate and had no intention of giving up. She remained close by calling her challenge.

Finally, our female gave up trying to dig and flew at the stranger, trying to drive her away. This didn't work either and soon the fight escalated so that they were actually trying to harm each other. At this point in the year any injury means it is unlikely that a female will be able to support young or reproduce that season. For a kingfisher this is a disaster, so they will do anything they can to avoid physical contact or injury. But they had been warning and whistling and showing aggressive postures to each other for several hours and it had done no good. Now it was a fight to the death.

Quite quickly the two females fell into the water, trying to drown each other. To resort to this is very rare and Charlie has only ever seen it once on the river. To our knowledge no one had ever filmed it, and his footage showed an incredible sight. Kingfishers when wet look even smaller and don't have the same grace as when they are fishing; they flap about on the surface of the water like dying fish. Each bird had hold of the top of one of the other's wings with its razor-sharp beak, and they were spinning slowly in the current. Their feathers reflected the sunlight bright blue and were spread out in the water as though someone had dropped a Victorian fan into the river. In this position, inch by inch, they moved downriver, occasionally adjusting their grips and trying to lever the other bird under the water. Neither was winning and neither was losing, and by now Charlie had completely lost track of which one was our bird.

They released their holds and flew up out of the water, each resting a moment, one on a rock, one on a branch. But they had resolved nothing yet, and it was only seconds before one had lunged at the other and they were both in the river again. They seemed evenly matched, but little by little the fighting tailed off as exhaustion took over, and they looked like they were holding on to keep afloat rather than drown each other. They were hardly moving and resembled two exotic petals which had collided as they drifted downriver. Their progress was then halted by some twigs that had been washed against the bank in the last flood. Once again those sharp beaks stabbed, but now there was very little energy left and it was beginning to seem as though they might both die. Charlie, watching through the viewfinder, was captivated, this was such a rare and emotional spectacle, and he was even able to keep them in focus!

Then came the unexpected. Suddenly a mink exploded into view from the bank. Quick as a flash it grabbed one of the kingfishers and was gone again. Charlie had no idea the mink was even there but still he managed to keep the shot in focus. It must have been alerted by the kingfishers' whistling and, like Charlie, had kept itself perfectly hidden on the bank waiting for the right moment. The mink carried the kingfisher up the bank, killed her and stashed her away for later. After the long struggle she was finished off in seconds. Then the mink returned for the other bird, but she had completely disappeared. It was only when we studied the footage that we realised how she had escaped. As the mink struck she had summoned the energy from somewhere and, released from the deadlock, had managed to dive. She had swum for a few feet under the surface and then flown up onto a perch.

And that perch, just beside the nest site, was where Charlie discovered her with the camera a few minutes later, wobbling and shaking. Her mate also found her, found her and then cemented their union; it was as though he wasn't going to leave anything else to chance. She didn't have the strength to resist and I'm surprised she didn't fall off her perch. However, it did prove that the survivor was our bird. She had successfully defended her territory, and in the fight of a lifetime had emerged exhausted but unscathed. Now she was ready to begin the season and reproduce.

It would have been a good day for Charlie to have got kingfishers mating, but to have witnessed such drama and captured it on film was what we had been dreaming of. Now I understood why he was so ecstatic.

Crufts was won by a poodle and we all said our goodbyes over glasses of champagne.

Once I was back from Birmingham we still had a few days to wait. Shooting on film is not like shooting on tape; you can't just watch it back. It's like taking photos. You have to get it developed and then transferred onto tape before you can look at it. Charlie's mind had gone blank. He couldn't remember if he had really got it or not – he thought he had, but you can never be sure if your shots will come out as well as you hoped. It seems so primitive, all this waiting and hoping, but it is part of the process.

The transfer turned out to be fantastic, there were no faults on the film or the camera, and the shots were all in focus. It was our first roll of film for *My Halcyon River* and the best thing Charlie had ever shot. They even showed it at the big Natural History Unit meeting. We had a long way to go but it seemed as though the gods were with us. So far so good.

Chapter Eight
BUBBLES–AU–VENT

....

An old otter hunter's cry

It wasn't long into our first season of filming that we realised Charlie needed help. Let me explain. Some nights were otter nights. They were the nights when we suspected an otter would be on the river. The female was moving around with her babies now; we could tell from the amount of spraint and the tracks up and down outside the house. It was very exciting, but we had yet to see her and yet to film her again. Charlie would rig the cameras at around dusk and then wait up all night; I would watch TV and then tuck myself up in bed and babysit.

After a night sitting by the river Charlie would come home as we were waking up, then we would all have breakfast and he would start work again – recharging the batteries for the camera and recorders, sorting out the kit, cleaning and rearranging, soldering and creating new recorders, repairing any that had got too close to the water, trying to lay his hands on some new lights and working on a good infra-red system to light the otters in the dark without them knowing. He might take a trip to the stores in Bristol to get some equipment or film, or he might take the dogs or just his waders and wade up and down the river all morning, looking for spraint or any other signs that would show which way the otters had gone. Then he would start to film again because there was lots of daytime stuff still going on which needed to be shot. There was no time to lose. Bluebells were in full flower, bright yellow primroses were smiling from every corner and babies were already abounding.

Ducklings were appearing every day and then sadly disappearing when they got eaten. We saw the dipper baby out of his nest for the first time and christened him Baby David. He stood on the edge of the weir bobbing and begging. The dippers are lovely to watch but you don't see them hanging around much. They aren't flash birds – they have plain brown bodies with white chest bibs and are about the size of a blackbird – but what makes them lovely to watch is their behaviour. They can swim and hunt underwater and are specially adapted to do that, with dense waterproof plumage to insulate them. They are exceptionally strong and have nasal flaps which close as they go under the water. There they either 'fly', using their wings as flippers like penguins to keep them under the surface, or they walk along the river bed, using strong legs, toes and claws to grip and their beaks to unearth and eat insects and tiny fish.

What makes them instantly recognisable is the bib. I have read that this is also an adaptation for underwater hunting, although the theory does sound a little far-fetched. The idea is that the white bib acts as a light reflector on the dark bottom of the river and so helps them find prey. Whether or not the bib was helping, our dipper family were hard at work feeding their fat, fledged baby and trying to encourage him to dive for his own insects.

We had two baby rabbits playing chase on the driveway and then havoc with the flower bed, but as we suspected that was where the mother had nested to have them and we left them alone for a while and just enjoyed watching their games. They were unashamedly fluffy and cute, and we didn't need to film them because they didn't have a part to play in a film about a river. They liked to play on the small humpback bridge that leads over a small stream into the kitchen garden. The game was a cross between king of the castle and tag with the pretender to the throne rushing up the bridge in mock attack. When they touched both would leap high in the air as if they had just had an electric shock, twisting as they fell. It is only a small bridge and I was terrified that they would end up in the mud at the bottom of the stream, but they would always land safely and tear off to play hide-and-seek underneath the shrubs. It was great fun to watch them but they were a real distraction from my attempts to get a schedule done. If you want to work on anything for longer than five minutes in this house you really need to shut the curtains.

Dusk would come around quickly and it would be time for Charlie to start setting up for the otters again. It was a very busy time, but there was one thing lacking in his personal schedule. I did say something, in fact I tried delicately broaching the subject a few times, but whenever I mentioned that he might think about getting some sleep, or remarked that his body would soon give out if he didn't get at least a few hours, or that perhaps he couldn't think straight because he was sleep-deprived, I got my head bitten off.

In the end I merely waited until he fell over. When he did, he looked up and said, 'I need help.' So we ended up with Jamie who, like a fool, agreed to work with us. Charlie had met Jamie on a different job some time before and thought he would make a great member of our team. His ambition was to become a wildlife cameraman, so we thought that a job as Charlie's assistant might be useful experience. We told him it would be a fantastic opportunity. He is enthusiastic and knowledgeable, hard-working, completely trustworthy and prepared to stay up all night looking for otters. And there is more.

At first I was quite apprehensive about getting an assistant. I knew we needed one but I was nervous about the intrusion. Because the location was our home and the whole project so personal, whoever joined us would become almost part of the family. They would have access to the house day and night, they would need to be there when we weren't and would have to be very discreet. We would see a lot of them and they would meet us in our pyjamas. What if they got on my nerves? However, my concerns evaporated when I clapped eyes on Jamie.

He doesn't have a clue, but when Jamie is on night duty at our house I tend to be very popular with my single female friends. Suddenly they want to come round for a few drinks or supper. They casually lounge around in the kitchen, giggling loudly at crappy jokes and pulling their hair under their chins in an alluring manner. It's not me they want to see or even our incredible baby. It's Jamie.

Jamie is gorgeous, although it would be wise of me to remind you at this point of how gorgeous Charlie is as well, just so we don't forget! Jamie looks like a supermodel, not a waif-like one, and, let's face it, although they look good in magazines they couldn't defend you against a puff of wind. No, Jamie looks like a red-blooded supermodel. He has nice broad shoulders and a square-jawed but friendly face, he is softly spoken and verging on shy. He is also much too young for most of my single girlfriends and completely in love with his equally gorgeous, charming and bright girlfriend, which is wonderful but only serves to make him even more attractive to the opposite sex.

He would also be spending quite a bit of time at the manor, as this was one of our major locations and currently a focus of otter activity, so we introduced Jamie to Richard and Stephanie. He would be parking in their driveway at all hours and sitting beside their river and we didn't want them setting the dogs on him. It was strange, but even when Richard was away for a few days, Stephanie didn't mind in the slightest that Jamie might turn up in the middle of the night; in fact she seemed quite pleased.

We were delighted that Jamie was free to start work straight away and that finally Charlie could stop panicking and get some sleep, although when they knew there was an otter on the river the two would stay up all night. Jamie would be posted just upstream from the holt where we thought the otter was and Charlie would be downriver. That way, whichever direction the otters turned, we would know where they were heading and have a good chance of filming them. We got lucky just a few nights after Jamie had joined us when he got closer to an otter than we might have imagined.

It was late May, Jamie was at the manor on duty, and by sundown he had set up a camera. In those early days we didn't know how much the otters would tolerate, but we did know that they were very shy and so didn't want to push them too hard. We set up cameras on remote systems, plugged them into a monitor with a recorder and then retreated to a distance, often using a car as a hide. That way you could watch the whole river through the camera but not be too intrusive. If an otter came, you could simply press 'record' on the monitor and then track its movement with the remote. The drawback was that the remote shots weren't ideal. The results looked a bit like the pictures from those security cameras that follow you in underground car parks and make you feel guilty about doing nothing.

Nonetheless this was a good starting point. At this stage we felt that any old shot of an otter was a good shot. We also had our theory that if a shot was a bit ropey then viewers would realise how special it was to watch wild otters on a British river at night. We reminded ourselves of this in those embryonic brainstorming sessions round the kitchen table by repeating such mantras as: 'Most natural history films are very polished,' and 'Snug and safe in your armchair with your glass of wine you forget what it is like actually to be on top of the rainforest canopy,' and 'Sometimes we take these incredible sights for granted because the shots are so slick.' We wanted people to get the same sense of privilege that we experienced when they saw the film.

So, Jamie was lucky enough to be spending his Friday night upriver from the house at the manor, just a hundred yards from one of the largest otter holts we knew of, sitting in his girlfriend's clapped-out old Peugeot, watching his monitor. It sounds like an easy job, but you can't have the radio on or the engine running, which means no heater, and even in May it gets a little chilly by the early hours. You can't read a book because you have to stare constantly at the monitor searching for any little ripple that might signal that an otter is coming, which means that you can't even pick your nails. You have to be always on the alert and thankfully Jamie is. He had seen the signs and pressed 'record'.

He soon saw a large dog otter making his way upriver towards the car and the camera. The otter had most probably just got up and now he was hungry. You could tell that because he was going slowly, intent on fishing. As we watched the shot later, while making toast in the kitchen, we were all delighted. The otter was clearly visible as he moved through the water and was not only large in stature, but as he got nearer to the camera we also realised that he was a little on the chunky side. This was an otter with several layers of fat to keep out the cold; the fishing must be good on our river.

We wondered what he would do when he saw the camera on its tripod. Would he immediately swim away or would he hang around at a distance for a closer look? He was swimming right up through the central part of the river, his head under the water most of the time as he hunted for food, turning over rocks with his front feet to find the fish hiding underneath and scouting around for a nice juicy trout or eel. The camera was positioned ahead of him on a little gravel beach which had been created by the river in full flood as it flowed around a bend. Jamie used the remote to follow the otter's progress. As the shot moved with him we expected a reaction to the sound of the camera on its remote head, but there was none. He didn't seem to hear it. Perhaps it was the noise of the river, but he completely ignored the strange metal contraption on the beach. He got closer and closer; we held our breaths. This first encounter would teach us a lot: it would indicate how close we were going to get to the otters, give us an idea of what they would tolerate and show us how easy our job was going to be over the next year.

The fat dog otter just carried on coming; the camera might as well not have been there. The nearer he came the stranger the angle of the camera got, until eventually it was pointing straight down at the ground. We gave a cheer as the otter bumped into the tripod. The shot was completely unusable – at such an odd angle that it made you feel strange to watch it – but we had learned a lot about the sensitivity of the otter. As he wandered back into the river and went off into the darkness, we remembered that when he first came past the house we had had all the outside lights on, the windows open and the stereo blaring. Perhaps these otters were bombproof.

In one way this lack of reaction came as no surprise. Although they are famous for being wary and secretive, otters already have plenty to deal with on this river. They negotiate roads, villages and factories; they go past noisy pubs and underneath bridges on which teenagers stand around hitting each other and smoking; they go past a sewage works and through countless gardens with barking dogs; they dodge cars as they cross roads and of course cope with all the natural obstacles which the river puts in their way. They have had to adapt in order to return to modern rivers and make their homes there.

Jamie had had a brilliantly successful night and we were all very pleased. It might not be great but we had a proper shot of an otter and we had a better idea of how our otter liked being filmed. The next step would be to introduce him to his cameramen.

By late spring the wagtails had lost their nest. In the darkness of the night it had been raided by a rat, who had eaten all the eggs, and they had had to start again. In order to coordinate the arrival of their chicks with the mayfly hatch they had to work extremely quickly and they did. The pair built a new nest tucked into the back of a pile of logs in a recess in the patio. It certainly looked snug if a little near to the ground. They soon had chicks however and spent ages wandering in odd directions over the patio, beaks stuffed with insects, trying to convince us that they had absolutely no interest in, or reason for going to, the log pile at all. Then, when they thought we weren't looking, they would make a dash for it and feed the ravenous chicks.

The rats seemed to be doing well. A female mallard had also nested close to the house, right beside the bridge, but her nest was too exposed and our hopes of watching ducklings hatch out right beside the house were crushed when we woke up one morning to find the nest abandoned, the mother gone and all the eggs smashed and eaten. Mink and otters will also take young birds and eggs, and so our nesting moorhen, ducks and wagtails had a scary time of it at night. There is little they can do to defend the eggs apart from sit on them. We resolved that at some point when we had time we would set up some cameras and try to catch this hidden side of the river's nightlife for the film.

By early June Fred had learned a word, and in the manner of small children repeated it endlessly. Predictably it was 'otta'. Further up the river at the farm the first kingfishers had already fledged, and the yellow flag iris was out up and down the banks to inform us that summer was now official.

Otter filming progressed slowly, but there were plenty of otters around and Charlie and Jamie got into a rhythm. It was a juggling act between the wonderful summer days and the glorious otter nights and in the end we realised that we would have to concentrate on one thing and that would have to be the otters. After all, there was no guarantee that they would be here next summer and we would be fools to let this opportunity pass us by. It was so rare to have the chance to study a mother and cubs, it would make our film unique. We decided to shut our eyes to the beauty of the summer days and pray that next year would be just as good.

Chapter Nine
CHANGING COURSE

....

An abrupt change of fortune

Charlie has never been an aftershave kind of a man. However, he does, I am relieved to say, wear deodorant. I have to confess I have never really noticed what flavour it is but there are others who have spent nights with Charlie who are evidently more observant than I.

Filming was progressing nicely. Slowly, cautiously and gently we were moving on from remote cameras. These were still in use and getting some really good shots – it was easy to rig them in all sorts of positions where it might be difficult for Charlie and Jamie to stay for any length of time, hanging off bridges or on narrow banks – but when he knew in which direction the otters were moving Charlie was trying to get close with a hand-held camera. In constant phone contact, they would spend the night judging the otters' speed and direction and driving down the lanes to different spots on the river to try to pick them up as they travelled through. Some nights it was easy: the otters would be in no rush and would fish and play as they went, rolling stones, toying with freshwater mussels and allowing themselves to be distracted. Other nights mother would lead the way, heading for some mystery fishing ground and determined to get there before the cubs got too tired. Those nights it was more difficult to keep track of them. They were never as fast as the dog otter on his own and the mother would have to keep waiting for her cubs, but there was little to announce their arrival at points down the river and they would be through so quickly the boys needed to be alert to spot them.

Fortunately, the mother tolerated Charlie and often just ignored him as though he were one of the trees, another dark shape on the bank. Little by little, night by night, he learned about them and they learned about him. Jamie and Charlie made a good team, and they needed to be; it was hard work – long nights in wet and grotty conditions with a low success rate. They became adept at communicating with few words, and their tiredness reduced the level of their humour to that of a pair of five-year-olds. Jamie's intellect became the butt of Charlie's jokes and name-calling, while Charlie's digestive system was the source of much amusement for both of them. However, they took the work very seriously. Once they knew where the otters were the banter stopped.

I knew we were winning the morning Charlie came home grinning from ear to ear. Eagerly I sat down at the kitchen table ready for the morning's viewing but there was no usable shot. 'They were too close to film,' he said, his eyes shining. 'The male cub touched me on the leg, the female bumped into the tripod. They came right up at me. I couldn't get a focus on them, they were too close.' We looked at the shot. It was incredible to watch, as though he hadn't been there but for the fact that as the cubs bumped into Charlie and the tripod the mother began whistling them away. It was for her, if not for them, a little too close for comfort.

But then, overnight, after all our careful work, the otters changed their behaviour completely. Now they wanted nothing to do with Charlie at all. At first he thought he had lost his knack for sighting the otters and was simply missing them. As usual, Jamie would film them on the remote camera and then phone Charlie. They would work out where the otters were heading and at what speed and Charlie would position himself on the river. But the otters didn't show. He was flummoxed.

After this had gone on for a week the cracks began to show; no shots and no sleep were not a good combination. Charlie moped around the house with drooping shoulders and moaned a lot about how boring his life was.

'We never have adventures any more. I used to go travelling all over the world,' he whined.

'What do you think this programme is, a walk in the park? This is the chance of a lifetime. Some people would give a limb for this kind of adventure,' I would retaliate. 'And half the time when you go away you write to me saying you want to come home.'

He then retreated to the kitchen and improved his mood by immersing himself in cooking. A fantastic bonus that came with Charlie is that he finds cooking therapeutic – which is lucky because when he is grumpy it reminds me that he is, after all, the man of my dreams. It was Thai prawn curry that night, his favourite and mine. Every cloud has a silver lining.

Later that evening, after clearing the chilli smoke from the kitchen and squabbling over who had had the most prawns, we discussed the problem in a constructive way. We agreed to investigate thoroughly the next day to see if we could get to the bottom of it.

After a good night's sleep Charlie resembled a human being again and we ventured forth to find out where the otters were. What we discovered was very interesting: the cold light of day revealed prints which showed that, rather than pass Charlie, our mum was leading her babies out of the river into the surrounding fields and right round him, entering the water again further downriver where it was safe.

It seemed as though we were back to square one. Yet what

was so strange was that the otters had previously become so habituated to Charlie's presence. I am ashamed to say that I merely shrugged my shoulders and was of no use at all. I could only think that Charlie must have offended them in some way. I did tentatively suggest this but realised immediately that it was the wrong thing to say. Back at home he stomped off upstairs to have a hot bath and wash some of the river water and frustration away.

Half an hour later there was a lot of yelling from the bathroom. I made my way upstairs wondering what had happened. Had he got his toe stuck in the tap? Had we run out of toilet roll? Was there an otter in the bath? Had he suddenly realised how big the hole in the ozone layer was? In the bathroom I waved my way through a fog of deodorant. What with that and the chilli smoke from the night before, my lungs were beginning to feel the strain.

'Smell this.'

'I can't do anything but. This isn't good for you or the ozone layer, you know.'

'No, smell it. Really smell it.'

'I can. It's horrible.'

'Thanks.'

'Well, it is. It's too much.'

'Does it smell familiar?'

'I don't know. It's a bit like oranges. Is this the one you normally wear?'

'Hah! Fantastic!' He had lost it. I was going to have to call someone. The fog in the bathroom was becoming thicker.

'What colour is the bottle normally?'

'I don't know, darling.' I couldn't see what he was getting at.

'I've changed my deodorant!' His grin was now inane, the glint in his eyes positively insane. All I could think of was how to get him out of the bathroom and away from the razors. My mind began to buzz, my throat was dry from the deodorant powder blocking up the pores, and all I could smell was musky oranges. I couldn't swallow any more. And then, finally, I caught up.

'You've changed your deodorant!' My inane grin matched his. 'Surely that couldn't be it.'

'I bet you any money it is. I'm not wearing any from now on.'

I'm probably the only girlfriend I know who would receive that kind of information with delight.

Sure enough, that night Charlie took care to shower again before he went out, he used no deodorant and the otters immediately returned to normal. They dawdled past him as he filmed them while standing in the river. Evidently they recognised his unadulterated smell as belonging to someone who meant no threat. They were probably grateful for his BO, it saved them a lot of time and trouble trekking across paddocks.

Meanwhile we were still keen to get our original shot, the one of the otter slipping out of the water just outside the house with all the lights on in the background. We thought it gave a great sense of how close otters can be to human dwellings with those inside having no idea of their presence. For me this is an important part of the magic of the natural world. I love the fact that if you look again you can see the exotic right under your nose. Every night we would set the shot up. The cables would trail over the patio and in the front door. The TV, with *EastEnders* on, would be surrounded by smaller monitors showing the outside of the house. We probably have the best security system in the world. Any burglar would be filmed in close-up, on a couple of wide shots and from several different angles. Compared to some of the close footage Charlie was getting in the water with the cubs this shot was unambitious, even basic, and yet still it eluded us.

Chapter Ten
LEARNING TO GO
WITH THE FLOW

....

Or, rowlocks, who needs them?

In September it was my birthday. I keep trying to forget how old I am but sadly can't quite manage it completely and this was my thirty-third. I'm not really sure what it says about my personality, but for my birthday that year I received from my lover ... Well, it started off as a surprise.

First thing that morning, I was told, 'You are not allowed to go outside.'

'Well, for how long?'

'Until I say.'

'Why?'

'I'm not saying.'

'What if I want to do some gardening on my birthday?'

'You can't. Go and get your coat. We are going on a magical mystery tour.'

I liked the sound of that so became instantly cooperative.

However, it turned out to be rather long and boring for a magical mystery tour with very little magic or mystery. So little that Fred gave up and fell asleep. We headed out into countryside I had never visited before, which became steadily more remote and very flat. I had no idea where we were going.

I spent most of the journey trying to guess what my present was. Were we perhaps going to some remote airport to catch a private jet to the south of France for lunch? But that wouldn't explain why I hadn't been allowed to go in the garden. Maybe it was a spa bath, an outside jacuzzi. I could just picture myself sipping a cocktail in a steaming bath beside the river while bubbles pummelled away my cellulite. But that wouldn't explain the magical mystery tour. It could be a puppy but then Bill would eat it. It might be— Charlie had begun to indicate, well the car had. We were still in the middle of nowhere but we were turning right into what looked like . . . Yes, it was a poultry farm.

Charlie looked at the expression on my face and began to laugh.

'We're not,' I said.

'We are. Happy birthday, darling.'

I began to laugh too. It was a great surprise. I had always loved chickens. Since we had moved to the house I had often talked longingly of chickens roaming the garden, of what a nice life they would have and of lovely fresh eggs. But what I like best about chickens is that they have always made me laugh, as have seagulls. I often take photos of seagulls, much to everyone else's bafflement. I think both species have great character and very purposeful personalities; they seem almost dogmatic in their behaviour, yet very self-important in their stance. Given a choice between seagulls and chickens, chickens have the edge. And now I could be entertained by my own chickens every day. I had been reading about this 'relaxing and fascinating' hobby in the bath, in the back of one of those countryside magazines, and Charlie must have picked up the article and realised that if ever anyone needed a relaxing hobby, it was me.

We pulled up outside a number of chicken runs full of chickens who all looked exactly the same. I now know that they were a cross-breed, Warrens, bred to be good, reliable layers, but at the time I only recognised them as the ones on the advert for Waitrose free-range eggs. It was a dreary day, bleak and grey, and the chickens didn't seem interested in us at all. Fred was still asleep and there was no one around. 'I did phone and let them know we were coming,' Charlie assured me, looking behind a few pens. Eventually a figure emerged from round the back of a distant Portakabin up the slope. The figure noticed us and descended the hill. She came over to inspect us.

'Frank's not here,' she announced.

'Oh!' replied Charlie, in a polite, interested way.

'I don't know when he'll be back. Gone to do some business.'

'Right.'

'So, will you wait?'

'Er, who is Frank?'

'They're his chickens.'

'Didn't I speak to you on the phone?'

'Mr Hamilton?'

'Yes. We've come to collect our chickens.'

'Frank's not here.'

'Right.'

This was getting us nowhere.

'Do you know which ones are our chickens?'

'Yes, those red ones. But I don't do chickens, Frank does them. I don't like them. I won't touch them. You'll have to wait for Frank.'

Slowly but surely I was beginning to understand the system: our lady dealt with the phone calls and Frank, whoever he was, dealt with the chickens. Despite working in a poultry farm our lady didn't 'do' chickens.

Now I came to think of it, neither did I. What were we doing? I didn't do chickens. I couldn't take these poor birds home. I had nowhere to keep them, nothing to feed them. What did chickens eat? I might notice if a foot fell off, but other than that I had no idea what illness might befall a chicken. I could spot the first sign of white spot in a goldfish but I hadn't a clue about keeping chickens.

And it wasn't just a question of keeping them. If you own chickens you have to handle them. They have beaks and scratchy feet. I was reverting to a complete townie. At first the idea of chickens pottering about the garden had been wonderful and appealing in a hazy, rural, Felicity Kendal kind of way, but suddenly I was full of doubt. Why didn't this lady pick them up? What did she know that I didn't? Doubt began to turn into panic. This was not one of Charlie's better ideas but what could I do? As each moment passed we were more and more committed to six chickens that we couldn't keep.

'Tell you what, I'll phone him.' Mercifully, our lady began the long trek back to the Portakabin. I was so grateful; now was our chance to escape.

Charlie was gazing at the chickens once more and smiling at the brilliance of his birthday surprise. I was about to burst his bubble. I plucked at his sleeve.

'Let's go.'

'What?'

'Let's go.'

'Why?'

'We can't have them, we can't keep chickens.' The panic was beginning to make my voice rise. 'Quick, before she comes back.'

'But I thought you wanted chickens?'

'I do.'

'Well, let's get some then. That's what we've come here for.'

'We can't.'

'For God's sake what is wrong with you? Why not?'

'I don't know what to do with them.'

'What do you mean?'

'I don't "do" chickens.'

'You don't have to "do" anything.'

'We haven't got anywhere to keep them. They need a little house.' That would decide it, we certainly couldn't just take them home and turn them out in the garden. They had to live somewhere.

'They've got a little house. I've restored the old hen house at the back of the garden and fenced off a whole area around it where they can live. That's why I wouldn't let you out there. I know plenty of people who have chickens. Don't worry, it's a doddle.'

I think Charlie had probably forgotten that the last time he said that, I was just going into labour. I hadn't. However, he had obviously gone to an awful lot of trouble for my birthday and there was no way he was going to change his mind now. I did feel much better knowing they would have a nice place to live. Even so, Charlie has a habit of making things seem a lot less complicated than they really are, so I quietly resolved to get a book, or several dozen. I would quickly become an expert chicken keeper.

Fred slept on.

There remained my most immediate concern.

'Do you know . . .?'

'Yes?'

'How to pick them up?'

'It's easy.'

'You can do that bit, then.'

Our lady returned. 'Can't get him.'

'Oh! Well, that's that then,' I said. 'Oh dear. We'll come back another time.'

'Right then, I'll get them,' said Charlie, rubbing his hands together. 'I've got some boxes in the back of the car.'

Slowly, a huge smile spread across our lady's face as she

looked Charlie's full height up and down. 'They're in there,' she told him, pointing to a tiny wooden hutch on legs. It can't have been more than three feet high and was raised about four feet off the ground. There were two sections to it, a small inside part and a larger outside run made of thick wire mesh. The floor of the outside consisted of mesh and planks, and was presumably designed to let the droppings fall through to the ground below and keep the interior clean. It would never hold an eleven-stone man. The only exit and entry point appeared to be a small door on the back of the interior section. Inside were six chickens with my name on them. Looking a little like an octopus squeezing himself backwards into a bottle and taking his legs slowly with him, Charlie climbed in.

Fred was still asleep.

Only moments passed before the door was flung open. I had the box ready and waiting and two chickens were shoved into it. By the time I had the lid down, the hutch door was shut and Charlie had dived back in for another round. I had expected a lot of squawking and swearing but the whole process was remarkably quiet.

I looked at our lady; she wouldn't even hold the box. 'I don't do chickens,' she reminded me. And indeed, it seemed to be true. She looked at them as though they were strange creatures at the zoo, truly remarkable things the likes of which she had never seen before. I wondered just how often she descended from her Portakabin on the hill.

The door opened. Two more chickens, looking a little shocked, went into the box. I shut the lid, the hutch door closed. This was easy.

You might think catching creatures in sheds is easy, but not always. I used to help out at St Tiggywinkle's wildlife hospital, where the deer were kept far away from the rest of the patients in peace and quiet in small garden sheds. Deer are very nervous and any noise will spook them. It is particularly important to disturb them as little as possible as they are notorious for having heart attacks. These sheds were cosy – thickly lined with straw and with infra-red lamps to make them really warm – and on cold winter afternoons I was often a little jealous. But however much you felt like snuggling up with Bambi, you didn't. There were very few people at the hospital who were willing to collect a deer for treatment, or who were allowed to.

By far the best deer handler was Les Stocker. Les, who had founded the hospital with his wife Sue, had learned through many years of bitter experience how the job should be done. He would go into the shed and shut the door behind him, there would be a few bumps and bangs and then he would call for us to open the door. When we did he would be holding a calm deer in a tight grip, ready for us to take it off for treatment.

Sadly, Les was not always there and when others had to step in, it was a different story. There would be banging, crashing and yelping (from handler not deer), the walls of the shed would literally bulge out as in a cartoon fight, and often there would be swearing. Usually the attempt was unsuccessful and the would-be handler would give up for fear of stressing the deer. The door would open and a red-faced young man (women were rarely stupid enough to attempt this) would appear with his clothes slashed from head to foot as if he were going to a punk concert. Deer have sharp hooves. We would creep quietly away, the deer would be left in peace and there would be much mumbling about how its health had improved.

Today, though, Charlie seemed to be enjoying himself in the small shed, and although it took a little longer to get the last two, I was soon placing a box of six happily clucking chickens in the back of the car. Charlie was a little flushed and flecked with chicken shit but maybe this new hobby wouldn't be so bad after all.

'Didn't you want a cock?'

'I beg your pardon?'

'Your cock. 'e's in that one.'

All credit to Charlie, who simply went a slightly deeper shade of red.

'Ah, yes.' The hand-rubbing wasn't quite so enthusiastic this time. The cockerel was huge. He was speckled black and white, tall and proud, with great spurs on the backs of his feet. His enclosure was a little smaller.

As it turned out Charlie had nothing to worry about. The door was only open an inch before the cockerel was out and running, showing off in front of all the hens and having a fine game of hide-and-seek with us. Then he spotted the tiny wood just beyond the pens and legged it. We followed as fast as we could, as did our lady, and I truly think that for her this was a defining moment. For as we grouped and regrouped in various formations for capture, as we sprinted and dodged, twisted and turned, it wasn't just the cockerel who was having fun. I think by the end of a rather spirited half-hour chase she was beginning to quite enjoy chickens and I wouldn't be surprised if she hasn't since taken up this relaxing hobby herself.

Finally we gave up and decided to settle for the girls only. We bought some feed and set off to introduce them to their new home.

Fred had missed the whole episode. He was still fast asleep.

As it turned out, how to actually hold chickens was the last of my worries because I spent most of my birthday trying to get them into their new house. I squeezed through hedges and under gates, combat-style; I tried various ways of untangling brambles from my hair while lying trapped in the hedge bordering the farmer's field at the top of our garden. At that point I allowed myself a short fantasy about what I'd be doing if my birthday present had been a hot tub after all.

Tina, my long-legged supermodelesque best friend, didn't desert me even in this darkest hour. She had taken a few days off to visit for my birthday and I knew that she had really preferred it when we spent birthdays shopping in New York or having expensive dinner parties, but in the manner of a true best friend she didn't comment once on my changed circumstances and became immensely practical: 'Perhaps if you moved forward an inch then backwards, you might get that bush to let go of your hair.'

Tina is endlessly loyal, the kind of best friend who features in *Mallory Towers*, the kind you never thought you'd be lucky enough to have. She is generous to a fault and consequently constantly broke. She can always find an excuse to open a bottle of bubbly and she can really talk. When I say talk I mean like no other talking you have ever heard. Faster than a racing commentator and twice as loud, she has found her niche in public relations. When she really gets going the volume continues to rise until I say, 'Tina, volume,' and she goes quiet again, but it only takes a few seconds for it to increase back to regular levels. Once, in a drunken moment, when Charlie was staring at her perplexed because he couldn't keep up, she confessed that sometimes her mouth goes too fast for her brain and she can't always remember what she just said.

Jeans on, make-up perfect and long, dark hair scraped back, like Posh Spice on the farm, she experimented with various shepherding devices, trying different noises and motions while I tried to extricate myself from the grasp of the hedge. None of our ideas worked, and at our lowest point we watched my birthday present wander off up a public footpath, miles away from their intended home. Covered in scratches and brambles, my T-shirt filthy and torn and sticks tangled in my hair, I had nettle stings down one cheek and all up my left arm and thorns stuck in the palms of my hands. She looked as if she was about to do a photoshoot for *Cosmo* in the countryside – cheeks slightly flushed, toned, tanned arms and perfectly white T-shirt. Eventually, Tina came up with her best idea yet. We left the chickens alone in favour of a bottle of white wine, and while our backs were turned, they all trooped into their lovely new house and lined up on their perch as good as gold.

I began to get close to my chickens in the days that followed. They were very tame and never pecked once. I named them all Flossy because they all looked exactly the same, apart from Edna, who had a middle toe missing. Five days after mine, it was Fred's birthday, and we had a big afternoon party. I was proudly showing off my chickens to a guest for about the sixth time that day when I nearly fell over. I had discovered an egg. In all the excitement I had truly forgotten that they lay eggs. There it was, a little brown gift which Fred ate for his birthday tea. I did notice Tina eyeing me with suspicion a couple of times. There are many prejudices to overcome when you start raving about chickens, however I do think that these six funny characters are actually one of the nicest birthday presents I have ever had.

Charlie had to go away. He had been asked to film a little-known shrew-like creature called the desman. Desmans live in the Pyrenees and Cantabrian mountains in northern Spain. These little animals are very cute with their big feet and long noses, but difficult to find, being nocturnal and only about the size of a rat. They were to feature in *The Life of Mammals*, a landmark series which the Natural History Unit was making with David Attenborough, as part of the insectivore programme. It was a short shoot and a quiet time of year, so we decided that a different job and a change of scene would not harm our film and would do Charlie the world of good. So, along with Sean the producer and seventeen cases of equipment, he happily jetted off to Spain the day after Fred's first birthday. The trip turned out to be a lot more difficult than either of them had anticipated, and filming secretive animals in the dead of night wasn't quite the change of scene Charlie had expected.

However, while they were away I had a great time. That September was the warmest on record and autumn was postponed for a while. Warm evenings are particularly nice in our house when you have a lot of washing to do, which is every day when you have a baby. This is because our laundry is up the garden in an outhouse. It isn't very convenient in the dead of winter when your smalls get frozen on the way from the tumble-dryer, but there is no room in the house so that's the way it is.

One evening I was floating back from dinner at my in-laws (my father-in-law is very generous with the wine) when I remembered that I had stripped the bed and washed all the bedding. I would have to get it out of the tumble-dryer and make up the bed before turning in. So, with Fred on one arm and the laundry basket under the other, I headed up the dark garden path to the laundry, followed by Bill. As I was fumbling about in the blackness trying to untangle Fred from a damp duvet cover I heard an unmistakable noise. I grabbed the dog, Fred grabbed the duvet cover and we all crouched down on the floor. It was the whistling of mother and cubs. It is the kind of noise you often think you hear, but when you really hear it you know you have. It is very distinctive: high-pitched and insistent. But what made me react in such a melodramatic fashion was that it seemed so close, much closer than the river.

Below the outhouse grandly known as the laundry is a pond, the last remains of the mill-race that used to run from our cottage down to the mill to turn the wheel. I was sure that this was where the whistling was coming from. They might be fishing or just playing but I didn't want to disturb them. We hadn't realised that they used this pond, and if they decided to do so a lot then it would be a great place for us to film them. One whiff of dog, yell of baby or scent of spring freshness from my newly washed bed-linen and they would be off. If I closed the door they would hear me. I resolved to run for it. I put the baby in the basket with the laundry and shushed gently at him, then with one finger looped round the dog's collar, half strangling him, I loped like some wonky old harridan down the path as rapidly as possible. In the house I shut the door, turned the river lights on and the house lights off, sat dog and baby down in front of the window and told them to shush again. I could tell from their faces that they both thought I had gone funny.

But my caution was rewarded. Minutes later the otters appeared, the whole family together, mother and two cubs. They were relaxed and in no rush, so my presence in the outhouse hadn't bothered them. They played in the river just outside the house for a while, the cubs rolling and turning on the surface and then diving. Because of the outside lights I could even see the bubbles rise from their mouths in little trails. What amazed me was how much they had grown. Although I had seen them in the flesh less often than Charlie and Jamie, I had watched them develop over the summer on the various tapes that had been triumphantly brought back home and played over breakfast in the kitchen.

It was easy to tell the male cub now. He was by far the bigger of the two, almost as big as his mother with a broad head. The female cub's head was shaped more like a seal's, her nose was less blunt, her features more pointed, and being smaller she seemed much more delicate. Her behaviour was also more reserved; she was less busy. Soon the male decided to get out of the water. He ran along the weir wall and then climbed onto the sloping concrete buttress which supports the bridge on the opposite side of the sluice. He paused there for half a minute and I got a good sight of him, as did Fred and Bill. He looked as though he could already survive alone. Big and muscular, he had developed a powerful tail, thick and long and tapering. However, he was probably not yet mature enough to go off into the big, wild world without his mother. To my delight he climbed onto the very bridge I had just crossed, where he scampered backwards and forwards for a while before plopping into the dark water on the other side where I could no longer see him.

As the ripples from his dive appeared under the bridge and spread back along the river I looked for his sister and found her just underneath the window. She was floating, and it seemed as though she was looking directly at me. Bill had begun to whine now that he had cottoned on to what we were watching, and I wondered if she was looking up because she could hear him. The fact that she seemed to be looking at me with just one eye added to the feeling that in contrast to her confident brother she was somehow shy and demure. I was sure she couldn't see me – there were lights between us and they pointed towards the river with me behind them – and yet she seemed focused on me. But then she was gone. Her mother and brother were moving up the river without her and she dived after them leaving only a trail of bubbles in her wake.

Neither Bill nor Fred understood my excitement. I found I was shaking, and although he was in Spain I left Charlie a long, babbling message on his answerphone, knowing that he would understand. To see animals like this from your home never ceases to be special. Each time I see the otters I am left with snapshots, pictures in my head. The moments pass so quickly it is as if you can't quite record them. I could look back at that evening as one of my best sightings of our otters so far, and the two mental pictures, one of him on the buttress and the other of her gazing up from the water in one-eyed wonderment, will always stay with me, even if they were never captured on the film.

When Charlie returned my call, he too was babbling. They had spent hours in the darkness out on the river trying to film the reluctant desmans in the Spanish mountains. For the fifth night they had had no sign, but just as they were giving up hope, he had turned to see – what else? – an otter cub whistling for his mother, who must have slipped past Charlie in the darkness to wait for him. It seemed to me like an awfully long way to go just to see otters.

Chapter Eleven
TROUT RISING

....

And disappearing

Autumn eventually arrived but not until the end of November. I was still harvesting chillies from the greenhouse and very small aubergines, which had suffered from a lack of water but tasted wonderful. We had a kitchen full of blackberries but our apple tree had been raided and our pear tree failed to come up with the goods so there was no apple and blackberry pie to sink to the bottom of our bellies. Instead I made more jam. I was getting better; this year you could spread it.

We spent one very interesting morning filming leaves. It was hard to believe we were seeing leaves falling off trees against a bright blue sky in November. Was this global warming or just good luck? Like September, October had been the warmest on record. Balmy day followed balmy day; there was not even a frost at night and so the leaves had not had the final push to let go and just kept clinging on. The delay had helped us no end by giving us much more time to film this particularly beautiful season.

In my role as producer, I had been auditioning leaves for weeks; they had been queuing up at the door in long lines. Joking aside, there had been hints that I did choose one particular leaf with rather too much care, and took too much time about it, considering that it would only be in the film for a few seconds. But this leaf seemed to me to be such a thing of beauty that I became quite attached to it. After we had filmed it falling off the tree, which involved a lot of subtle branch-shaking on my part and a lot of frustration on Charlie's, I kept rescuing it from the river so that we could do close-ups.

I can't tell you the number of times I have nearly fallen in that bloody river. Everything is so slippery that no footwear on this planet can give you a good grip. I wouldn't mind but wildlife cameramen seem to be as sure-footed as goats.

Charlie and I spent a happy morning arguing gently and exchanging insults, paddling, slipping, swearing and filming, and got the shots we were after. It was just as well, because at lunchtime we had to give up as the sun went in and the leaves went from looking like drops of liquid gold to resembling dull, dead leaves. So we spent the afternoon drinking tea and arguing over the script instead. Charlie was tired and irritable because he had been up all night trying to film the otters. I, on the other hand, behaved impeccably.

That night the temperature began to drop. You could feel a new chill in the air, the curtains were drawn around the house, and I lit a fire. Charlie began the evening resolved to get at least one shot of the otters but had a bad start when the otters went past the house at around quarter to seven. The clocks had changed, and although as far as we knew the otters had no idea of this, they had changed their habits too. We suspected that they would be leaving their holts earlier because it was now dark earlier, but they had been coming past us later and later and it was difficult to predict exactly where they might be and when. Jamie was on duty further upriver near the holt at the manor, and when we heard the otters go past we realised he was sitting outside the wrong holt. And because we hadn't been expecting them, we had no equipment ready and so missed a shot too. We decided to try and get ahead of them. We had some idea of where they had rested through the day – between our house and the manor – and from that and the time we could work out how fast they were heading downriver. The otters were going at a fair speed. They may well have been heading for new hunting grounds. The cubs were getting quite big now and they would need to keep finding new sources of food.

We had suspected for a while that they were using the carp ponds but had not managed to get a shot of them there. The ponds are also used by the local fishing club, and we live in dread of the fishermen finding out that there are otters around; they probably wouldn't be too pleased to learn they have competition. Now that we had lost the chance to catch the otters outside the house it was worth trying to beat them to the carp ponds.

Charlie had just finished his dinner so at least he had a full belly for a long stint beside the ponds. I didn't envy him as he put on his thermals to venture out into the freezing night. The otters normally took around two hours to get there so he loaded up his kit and drove to the ponds in the Land Rover. We were discovering that the key to filming these otters was being flexible and grateful to get what we could. He sat there for well over three hours. There was no sign of the otters – the carp were getting off lightly that night – so Charlie came home. He was cold, tired and fed up.

The next morning the alarm went at 4.30. I managed to prise open my eyelids to see a tall but drooping figure silhouetted against the curtains getting into cold muddy jeans and thermals, but it turned out to be well worth it. By the time Fred and I surfaced some hours later, Charlie was a very happy man. He had got his shot, but only just.

One of the most valuable lessons we'd learned over the previous year was that if you leave your post for ten minutes, you can guarantee that is when the otters will arrive. In the darkness of the early hours, Charlie had spent half an hour rigging another shot he had been trying to get for weeks. The plan was to catch the otters climbing up the weir, great if we could get it. The weir is at least ten feet high and you would assume the otters would circumnavigate it by getting out of the water, going up the steps and re-entering the river on the other side of the bridge. But we had witnessed them walking up the vertical wall of the weir as if it were no more than a slight incline, completely ignoring the water spilling down over the top. Charlie needed two cameras: one, rigged below the sluice, would capture the otters coming up the river and the other would pick them up as they climbed up and over the weir itself.

By 6.30 a.m., Charlie, understandably, needed to desert his post for a moment and put the kettle on. When he got back the otters were just leaving. But hurrah! He had learned his lessons well and had pressed 'record' before he left, so despite his craving for a cuppa we had fantastic shots of the otters coming up the river like dolphins and clambering up the face of the weir.

It was an important shot for the film as we didn't know how much longer the otters would be together. The cubs were almost grown and soon the mother would get fed up with supporting them and send them off to find their own lives.

But later that same morning, triumph almost turned to disaster.

We had agreed to look after Richard and Stephanie's trout pond while they were away in New Zealand. I always enjoy the walk to the manor and so do the dogs, so feeding the fish was a great way to exercise Bill and Honey and get some fresh air. And it never does any harm to wander up and down the river – it's a bit like catching up on the local gossip and you never know what you will see. This morning Charlie came with me to do some kingfisher spotting.

As I was throwing fish food pellets into the large artificial pond my gaze began to wander. The late autumn meant that the huge horse chestnut trees, with the sun behind them, were still glowing orange and only just losing their leaves. They looked wonderful. The pond was about three years old and I was thinking that it was now starting to look natural. The vegetation had grown over the bare patches around the edge where it had been dug and there were more bullrushes around this relatively new pond than I had ever seen in one place. They were just passing their best, some moulting a little and others falling over.

My gaze returned to the pond and I suddenly realised that nothing was happening. My pellets were floating on the top. There was no swish and plop of hungry brown trout, no gaping, searching mouths. No glimpses of long, brown, spotty bodies. No trout.

This was a worry. We had been feeding the trout every day for three weeks as promised, but now, the day before Richard and Stephanie were due home, there was nothing to feed. This was like looking after your best friend's goldfish while she is away and finding it floating on the surface of its bowl, but on a massive scale. We were talking at least sixty trout. A horrible feeling was developing in the pit of my stomach.

I called to Charlie, who was inspecting the riverbank, and I showed him how, when you threw in the pellets, nothing happened. 'Hmmm,' he said and immediately began looking about on the ground. I knew what he was looking for: otter spraint. Although we had had plenty of recent sightings of the mother and her cubs, the dog otter had been conspicuous by his absence.

My worst fears were confirmed when we found spraint, quite quickly. It was hard to miss, on a small rock at the edge of the pond. As if that weren't proof enough of the fate of the trout, there were bones in it and, most damning of all, a long, complete trout spine, stripped bare. There were spraints dotted around, some a few days old but one very fresh. Oh dear, what a quandary. On one hand this wasn't all bad news for us. The pond was a good, accessible place for us to film the dog otter as we could get the Land Rover really close with all the kit in it. On the other hand, there were no fish left and somebody would be facing a huge bill for lost trout.

The incident highlighted for me a major source of the conflict between otters and people – fish. Although we suspected it was only the male otter fishing in the pond the mother and cubs could be visiting too. Each adult otter needs around a kilo of fish per day and now that the cubs were almost full-grown that amounted to an awful lot of fish. If the three otters had been travelling together, in the last few months about sixty trout had been turned into otter spraint, a small pile of debris on a rock and muscle on two young otters.

Some humans think that they have rights over fish. The otter hunts from hunger but humans usually fish in the name of sport, yet they can be just as territorial over their stretch of river and their fishing rights. They will even accuse a heron, mink or otter of stealing. But surely no hungry otter can be blamed for hunting fish. We were extremely lucky that the owners of this pond felt exactly the same way. Far from mourning the loss of the trout, Richard and Stephanie considered themselves privileged to have otters fishing on their land.

Nevertheless we felt responsible, almost as if the otters were ours. Richard and Stephanie arrived back in a jovial mood. They had had a wonderful trip and were delighted to be home, but they were immediately concerned when they saw our glum faces. We grovelled and offered to buy new trout in return for forgiveness for the otters but Richard and Stephanie simply laughed. Relieved they weren't cross, we kept our promise to help restock the pond, even if our track record on trout left something to be desired.

However, there are many fishing clubs and carp pond owners who don't feel the same way, and I constantly fear that people will forget how precious the otters are and begin hunting them again. Having watched these cubs grow from tiny babies, the thought of large men wading and splashing up and down the river with baying hounds hunting otters and disturbing wildlife in the name of fishing rights makes me furious. It will always be the same though. Some humans don't respect the fact that animals have as much right to life as they do and certainly far greater rights over a few trout.

We had a great trout, living just outside the house. We called him Mr Smooth. He resided in a cave underneath the weir and the sluice and managed to stay there even when the water was foaming through the gates at its most powerful. Maybe it was the oxygen levels or the food supply, but it was obviously a good place to live. He was huge, at least six pounds, and you couldn't help but imagine that he was wise, too. Sometimes we threw him small lumps of bread from the bridge, our version of the doughballs in Pizza Express. He loved them. He seemed to come from nowhere, a huge mouth sucking down the ball, and then he always gave us a little show by flipping his huge body out of the water and splashing his tail before he disappeared back into his cave.

Back in the summer we had made the decision to concentrate on otters for as long as they were around. We had also taken on Jamie as an assistant. Jamie was invaluable, if always twenty minutes late. This was because he knew that Charlie was never on time and he would have to sit and wait for twenty minutes. Being late, therefore, made perfect sense. But the best thing about Jamie was the way he would work twelve-hour shifts in the cold or the rain and never complain. We could call on him day or night and he never murmured. Mind you, if he had moaned it would have been hard for him to make himself heard above Charlie's grumbling. He was like a rock, always there and we could rely on him one hundred per cent.

One of my major roles on this project was the catering and Jamie was always very appreciative of my pasta-bake suppers and flasks of coffee. He was also great for morale. The way the banter flowed of an early evening in our kitchen belied the fact that they were heading off into the night on yet another gruelling wild-otter chase. The humour remained the same. Up the stairs they would clump, gently taking the mickey out of each other's toilet habits or sexuality. Bump, bump, bump in the 'kit room', which now bore no resemblance at all to a guest room. A few insults would be exchanged about who had forgotten to put some batteries on charge and then clump, clump down the stairs again.

Jamie would invariably get sidetracked by Fred, holding him while I heated up purée or playing with his toys while Charlie went to the loo again. He was particularly good at pulling entertaining faces. Gentle and quiet and easy to have around, he put up with Charlie just as well as I did, and, to be honest, he probably spent more nights with him than I did. He had fitted in well and we knew we couldn't have kept our heads above water without him.

On the whole, the autumn was an optimistic time. So far, the gamble we had taken in concentrating our resources on night work and the otters seemed to have paid off. Late November however, went beyond our wildest otter expectations. We sighted so many of them we didn't know how to film them all and they even kept regular time, just to make our lives easy.

There was an unfortunate consequence to all this. We had been planning to film Mr Smooth. We weren't intending to do this until December when trout move up the river to begin spawning, but we had noticed lately that his presence was lacking. By the end of November it became apparent that so much otter activity had come at a price. Tragically, one of our major stars had been eaten before he had appeared in a single frame. He hadn't even made it to the cutting-room floor. If we were to film trout spawning in December we were going to have to make other plans.

November is when everything should be shutting down for the winter and we were anticipating a period of contemplation, viewing our tapes and carefully assessing our position so far. The otters had other ideas.

It was a month of firsts. We had our first sighting of an otter during the day, our first shots of otters using a holt and the unbelievable sight of four otters travelling together, the whole family. Oh yes, and there was the car stuck under the bridge.

As anyone who lives on or near a flood plain will testify, autumn is not only getting later, it is also getting wetter. Floods have become a predictable part of the end of summer. For some people this has meant misery and homelessness, but we have been very lucky. The waters rise but they have never touched the house. This means that even when the water is high and the noise of the river is roaring in our ears, we can still sleep at night.

Early November did bring floods in the area, although they were not severe. One night, however, in the tumultuous inky darkness, there was a strange noise outside. Briefly aroused from my slumber, I merely turned over and went back to sleep. The sound was a dull thud only just audible above the thunder of the water. We have become accustomed to seeing whole trees come down the river and past the windows. Usually they go over the weir but sometimes they get stuck in the open sluice, so the odd thump is nothing to get excited about. The river brings us all sorts of gifts. Footballs are frequent, as are tennis balls lost by unfortunate dogs on riverside walks. One year a stack of marijuana plants came floating down, presumably dumped in a hurry. (There was no sign of the otters that night and we could only imagine the kind of night they had had.) We have had wooden pallets and boats, a beautifully constructed raft, apples, lots of road signs, tyres and the usual litter that people enjoy throwing around to mark where they have been.

The dipper roosts under the bridge most nights and she must have been slightly more perturbed by the disturbance as when we got up in the morning, to our amazement, we discovered a car outside the house in the river, a red Peugeot 309. I don't know why, but the first thing I noticed was that it was fully taxed and the instruction manual was floating around inside. It had evidently sailed down the river and it was now stuck under our bridge.

It was a completely surreal sight which demanded that you go back into the house and come out again several times to appreciate it fully. The brown waters of the swollen river tumbled around it but didn't shift it at all; it was firmly jammed at a slight angle, the front bumper forced up against the bridge's metal outer edge. It must have either floated down backwards or accomplished a three-point turn before parking itself outside the house.

We did what every first-time film-maker does in an unusual situation: we got the camera out. I am ashamed to say that it was only after we had filmed it from many different angles and taken lots of photos that it dawned on us how potentially dangerous it was.

The most urgent question was whether it was leaking oil or petrol into the river. We presumed it wasn't; we couldn't see any sign of an oily film on the surface of the water and couldn't smell any petrol. There was a chance it had emptied itself of these horrible substances during the night, but if it had then the damage was already done although there was no indication that this might have happened. So we just had to hope for the best and get the car out of the river before the pounding waters took their toll.

We phoned the police. They weren't interested. All they wanted was the registration number of the vehicle so that they could trace the owner. They calmly explained to Charlie that the recovery of the vehicle was our responsibility since it was on our land. Charlie calmly explained back that it wasn't on our land, it was in the river, which we don't own because no one does. They still weren't interested. There was nothing they could do. We tried to explain about pollution but it was pointless. They never did trace the owner.

—Charlie put the phone down and moaned a lot and we both bit our lips. How the hell were we supposed to get the car out of the river? How much was this going to cost us? Suddenly the red Peugeot 309 didn't have the same comedy value as it had done five minutes earlier.

There was something else: the rain had stopped. If it didn't start again the water level in the river would drop. It only needed to go down by a couple of inches and the car would be freed from the bridge and float down the river straight into the sluice, where it would probably cause thousands of pounds' worth of damage. Any remaining oil or petrol would certainly spill on impact and the underwater current going through the sluice would probably sink it, making the car even more difficult to recover. Just to prove the law of sod the sun came out.

In the end it was the Environment Agency which came to our rescue. Although they were not obliged to help in any way they understood our concerns, particularly with otters on the river. With the help of the council they mobilised a rather large tractor and some rather large men to tow the car back up the river and out of the water into our paddock. It was a bigger job than anyone had expected. The car was full of water and had to be towed against the flow. Once they had a good grip of the car we lowered the river by opening the sluice, which made things a little easier, but the tractor had to strain to make even the slightest progress. It reminded me of an episode of *Bob the Builder*. The tractor seemed almost alive in its endeavour and the men around it were willing it on as it struggled through our marsh of a paddock. Everything had seized up in the car – there was no steering, the wheels wouldn't go round and it was full of silt – so once it was out of the water it had to be dragged. The whole operation was very smelly and very time-consuming. We did recover the log book from the glove compartment, but most of the writing had dissolved so it was impossible to work out who the owner was. I have often wondered what they put on their insurance claim form.

Our cubs were now easily as big as their mother and it was hard to tell if the young dog otter wasn't even a little bigger. They were by now accomplished hunters and fed themselves on their nightly forays. Charlie had come up with the great idea of putting sand spots up and down the river. This was at least an hour's work during the day but really paid off in the evening when we were trying to work out where the otters were starting out from. We spread builder's sand over the points where they entered and exited the river, and smoothed it each morning and evening. As long as there was no rain, it was easy to spot who had been where the night before and to calculate where they might be starting from the next day. This simple, clever technique had also been invaluable in the middle of the night when we were watching. If ripples on the water led you to believe that you might have missed an otter going by, a quick inspection of the sand would tell you for sure. The sands, like tea leaves in a cup, told their story, although they were open to interpretation. Our interpretation was that lately our young dog otter was going off on his own for at least part of the night.

At around this time the mother and cubs had got themselves into a little routine. They would be out and about at the same time early every evening and they either came past the house whistling away to each other or appeared up near the manor. It was not difficult to deduce that they must be staying in a holt in between. It was unusual for them to stay in the same area for so long, but we were delighted that they had decided to linger. We cancelled every social engagement and, despite miffed friends and family, determined to make the most of it. We didn't know how long they would carry on like this because we had no idea how long three practically fully grown otters would find food in this part of the river.

We discovered the holt one morning when the otters were travelling particularly slowly. It was not a location we would have put on our shortlist, an old drainage pipe by a bridge and a road. At the moment it was dry but there were times when water poured out of it, so it was not quite the snug, cosy home you might imagine otters curling up in.

The next evening we got a 'money shot', a shot that really made sense of otter behaviour. It was fabulous. As the afternoon drew to a close, we set up the cameras outside Bridge Holt, as we now called it, while the otters were still asleep. We wanted to get them coming out of the holt in as natural a way as possible. Jamie was put on duty to sit in the cold for the next few hours and film them as they emerged. Sure enough, at the usual time, the family roused themselves and came out.

The cubs appeared first. They took their time, seemed completely relaxed and showed no sign of even noticing our cameras on the opposite bank or the infra-red lights we had rigged. They lay in the entrance to the holt, framed by the pipe, groomed each other a little and played some kind of pat-a-cake game with their paws. You could clearly distinguish between the chunky features of the male and the smaller, more pointed face of the female. Minutes passed and gave us the chance to study them properly. They had grown even more than we had realised and put on a healthy layer of insulation. They rolled around a little as if they were lounging in the sun rather than anticipating a dip in an icy British river, and then to our delight seemed to get stuck in the mouth of the pipe. They had grown even more than they had realised too. The male, having decided it really was time to get going, tried to wriggle free of his sister and the pipe to get down the bank and into the water, but he was stuck like a cork in a bottle. His paws waggled uselessly because they could not get a grip on the side of the pipe – there were too many folds of flab in the way. It was a wonderful comedy moment and he looked distinctly embarrassed. In the end he used his sister as a lever and came shooting out onto the muddy bank, where he attempted to recover his dignity. His sister, meanwhile, began breathing again.

Then Mother joined her daughter in the holt entrance and they all made their way slowly down the bank and into the water. Three grown-up otters barely causing a ripple, the family headed off down the river to hunt and play for the night.

It was fantastic footage and that week just got better and better. Early the following evening when it was not quite dark, Charlie was doing the washing-up in the kitchen staring absent-mindedly out of the window contemplating his lot in life, which is what he always does when he washes up, when a large otter swam past. About thirty seconds later, three more otters cruised past, and they all went downriver together. This was the first time he had seen the four otters as a group but it was so early in the evening he wasn't yet ready to go. Rushing around the house like a lunatic, grabbing batteries, cables and cameras, he got the kit together in record time and drove downstream to Tipple Farm, where he set up and sat praying that he had judged it right and they weren't already ahead of him. The only way to tell is to wait, but if the otters have already been through then the longer you wait the further downriver they will be and the smaller your chances of finding them again. This was a unique opportunity. The whole family travelling together only happened on very rare occasions, and to miss it would be unforgivable.

Tonight his judgement was right. After ten minutes the dog otter showed up, closely followed by the other three. There was lots of squeaking. The cubs were anxious, remained close to their mother, and were still keeping their distance from Dad. He then spent some time fishing right in front of Charlie, taking his time and quite relaxed, sending very clear signals to the rest of the family that he was no threat. Yet the cubs and female still hung back, squeaking, nervous, peering. After a while they began to move again and they all swam by.

Our dog otter had shared the river with his family before but had always kept his distance. The four kept track of each other through their spraint, but we had never been sure how close they had come to meeting up. We were sure, however, that they had not been travelling and fishing together up till now.

Heart racing, praying that his luck would hold, Charlie packed up the kit again, chucked it all in the four-by-four and careered down the lanes to the next point where he thought he would stand a good chance of filming all four of them together, the carp lake near the village. This time he wasn't certain whether he had judged their speed correctly. They had taken off quite quickly, but the cubs weren't at all keen to follow the dog otter too closely. After an hour he decided that he must have missed them and should try another location further down the river. He stood up to begin packing away the kit. At that moment they turned up.

In just one hour so much had changed. Now they were all fishing together and much more comfortable with each other, although they were still spread out and stress levels were higher than normal. So were Charlie's. One look at him and the dog otter had his family off the lake and back to the river, where they swiftly moved on. Reluctant to push his luck, Charlie came back to the house and rigged up the remote cameras there. We had no idea whether they would return to Bridge Holt that night, as Mum and cubs had been doing for most of the month, or whether they would move on with the dog otter to pastures new, but after another night watching the water, the story unfolded.

At six o'clock in the morning, moving quickly and calling to each other constantly with whistles and squeaks, the cubs came home alone. Their mother must have gone on with the dog otter and, for the first time, had left her babies. The cubs came back up the river to the holt and went to bed on their own.

Once more we had so many questions buzzing through our heads. Would she return? Was this it? Had she now officially finished the business of rearing her cubs and left them to survive alone? I couldn't help but be a little concerned for them.

It was at one o'clock in the afternoon, and only by chance, that Charlie saw her return. What a dirty stop-out! He was working in the office in the house when he noticed great ripples on the millpond and dashed outside with the camera. This was another first, the first time we had seen an otter on the river during the day, and was completely unexpected. She was taking her time returning to the cubs in Bridge Holt. It had been their first night without her but she didn't seem concerned. She rolled about playfully in the water and even fished a little for a late breakfast. Charlie got a couple of shots of her but it was difficult because there were too many trees in the way.

We knew what this probably meant, and it was very exciting. She must have been with the dog otter all night, and the only reason for them to travel together was because they were mating. Otters in Shetland have been observed mating with cubs present, but those are coastal otters. So little is known of the river otter that its mating behaviour might well be different. The American otter has what is known as 'delayed implantation' which means that an otter can mate at any time, even if she has just given birth; that mating can result in a fertilised embryo, but the embryo will remain dormant until the right time of year and only develop in time for the female to give birth in spring. Little is known about whether this is the case for the British otter. By counting the months we estimated that this would be roughly the right time for our female to mate if she were to give birth in early spring, without delayed implantation. This would be at around the same time as this year's cubs had been born. But where would she choose to have them? Would she bring them back to this river or was it just the early floods that had forced her to bring them here? Suddenly we had lots of questions and although we were impatient to know the answers, we would simply have to wait.

Chapter Twelve
WATER LIKE A STONE

....

The big freeze

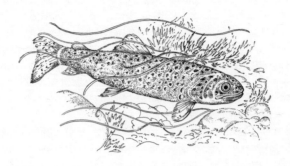

O wing to such a great otter month in November we were now running really late to prepare and film trout spawning, and December was our only chance, the only time they would spawn. This time next year the programme would be sitting on a shelf somewhere in the famous BBC Television Centre waiting to be broadcast. It was an intimidating thought. It would have music, titles and credits, stories, highs, lows, dramas, funny incidents – none of which we felt we had yet. I was finding it really hard to imagine that it would ever be finished, and every time I thought of the programme I panicked. So, in early December I decided to think about trout instead.

I had been reading up on trout. As a working mother I was beginning to learn that you take advantage of any scraps of spare time slung your way by the gods, and I did most of my trout research on the floor of first class on a train out of Paddington. The trip home from London is often delayed, and on this occasion my briefcase was full of notes and books from the library, everything I could find relating to trout. We had been shunted from one train to another and from one platform to another as each train, exhausted by the week, decided that it wasn't quite up to the journey. Although I had been early enough to get a seat on the first one, the subsequent moves ensured I quickly lost it, and while we waited another forty minutes for this particular train to work out whether it was up to the journey west, I decided to withdraw from the general commuter hysteria gathering in the carriage and ducked down between a suitcase and the bin. Here, on the beery carpet, I immersed myself in the world of the trout and escaped the world of the not so great western railway. The time was mine; at home I would have had to play diggers or, worse, robot wars. Here I could work.

There is a large amount of literature on the trout, mainly concerned with catching it, and much of that is about preferred fly types. However, there I found more than enough information useful for the programme.

You might think that not a lot happens when you are a trout, but you'd be wrong. Actually trout have a very interesting nightlife. They do their hunting at night, which might help to explain why otters are also up all night. Suddenly, when you think of trout as nocturnal hunters, the subject becomes more interesting. Also trout are extremely territorial. They find a good spot in the river and then fight to keep it. Trout the fighter – sounds even more interesting, eh? The size of a territory depends on food supply and the density of the trout population, and some territories are better than others. Presumably, the late Mr Smooth's spot is highly coveted. It is the perfect location for shelter from predators, incoming food supply and good oxygenation. I could see only one problem for the next occupant: what does he do at spawning time when he needs to leave his spot and venture forth?

The most exciting time of year for the trout is probably

autumn, which is when they begin to spawn. This can occur any time until mid-December, and captive trout will often carry on until mid-January. Trout take advantage of autumn floods. As the waters abate our male trout will begin to move upriver looking for a place to have a one-night stand with a hen. (Females are hens and males are cocks, but then we all knew that already.) Trout are quite particular when it comes to selecting a love nest; it has to be just right. The water should be quite shallow but deep to enough for the trout to swim. The substrate should be gravel, preferably pea-sized, because the hen swishes it with her tail until she has created a shallow depression in which to lay her eggs. Trout apparently also prefer the lower end of a pool, although I can't be sure why. Mr Smooth's successor will also go through a few hormonal changes. At spawning time the male's colour intensifies and he develops a hook on his lower jaw which increases in size with age. Again, none of the literature specifies what this is for. Maybe a big hook is a sign of virility, of good genetic make-up, maybe a big hook attracts the females.

As for our river, this is where it gets tricky, as it is hard to think of a place which has all the right conditions. There are plenty of shallow spots – where the river runs over the rocks up at the manor and further downstream from us, even towards the head of the river near the motorway. However, none of them seem to have a gravel substrate. In fact there is very little gravel on our river at all. So where do the trout go?

The other thing that puzzled me was the difficulty of moving up our river if you are a fish. The literature talks of trout passing-places. These exist where a river has been blocked – by a weir, for example – and are designed to allow trout to move upriver when they are spawning. However, I had seen no such thing on our river and we have four weirs: one at the factory, one outside our house, one at Valley Farm and one at Court House. So how on earth do our trout even start that incredible journey?

Take Mr Smooth's journey, for example (God rest his sole!).

The River

When he was alive, to get up the river and procreate he either had to ascend our weir, which is at least ten feet high, or go through the sluice gate. The sluice gate is not always open, for a start, and when it is the water thunders through it with such force that I could not imagine how any trout, even a huge one like Mr Smooth, could ever make it.

So, how on earth were we to film the trout spawning, autumn's major river event, when we had no idea of where they did it? We researched locations on other rivers, and discovered that on some, spawning sites are actually created for trout. But it didn't feel right using another river; it would have been cheating. We decided to prepare a tank just in case. If we kept the trout in a very large, natural tank we could monitor their behaviour and also film them from many different angles, which we couldn't do in the river. We could also show how their colour changes. So Jamie and Charlie set to. In a week they had built what amounted to a trout jacuzzi in the garden a few yards from the river. I call it a jacuzzi because, in order to mimic the strong flow of the river, we had to use jacuzzi jets.

First of all they built a large wooden frame, around five metres by three, then they went and had long discussions with the man in the local fibreglass shop. After much debate we decided that the best route forward would be a fibreglass tank with built-in glass panels. Because of the size this would be the strongest and safest option. Days and days were spent spreading and smoothing the foul-smelling mixture over the wooden shell. Meanwhile, Charlie's little brother Jeremy had been on the internet looking at every hot spa and make-your-own-jacuzzi site he could find. He had researched and ordered the right pumps, filters and nozzles for the job, and when they arrived that meant another day fitting and a little more money than anyone had anticipated. Another few days were spent filling the tank and waiting to see if it leaked. The cycle of emptying and fixing, filling and waiting seemed to last for ever.

Finally it was ready. We furnished it with rocks and vege-

tation from the river, and of course with water, and we left it
to stand for a while before the ceremonial turning-on of the
pumps turned it into the very picture of a fast-flowing well-
oxygenated river. And, late one winter's night, after a long
journey from a farm in the middle of the Cotswolds, our
brown trout arrived. They were shown into their new home
by torchlight. We were all relieved that we could start the real
work of observing their behaviour and filming them the next
day. But, later on that same winter's night, they left. This was
not the fault of the otters, as you might immediately suppose,
or even a mink. It was their own decision. We would not have
believed it if we hadn't seen it with our own eyes, but in the
morning, when we went to check on them, we were in time to
observe one last trout flapping his way down the bank towards
the river and, with one last tremendous effort, flicking himself
into the water and swimming off to spawn. We had placed the
tank near the river so that it would have a natural-looking back-
drop but the view from the windows had been too enticing
for the trout. They wanted out and they got out. All bar one
who seemed very happy, thanks very much, and decided to
stay. However he only decided to stay on one condition: that
he didn't have to come out from behind his rock and be
filmed. He didn't want a career in showbiz, just a quiet life.
Unfortunately, you can't really film just one trout mating.

I wouldn't mind so much if it wasn't for the effect on the
view from the kitchen. I have to look at the ugly, square
fibreglass-and-wood monstrosity every day because, of course,
no one can move it. The jacuzzi system sticks out from the
back as a reminder that this is no ordinary jacuzzi and certainly
not the inviting sort, with steam and bubbles, that you see in
the back of posh magazines. This is the slimy sort with an odd
collection of invertebrates at the bottom. All the children who
visit love to peer into its murky depths through the algae on
the glass, but the view is sullied and the trout mating sequence
never did make it into the film.

★

December was turning into a bit of a fallow period. The otter filming wasn't going very well for various reasons. Charlie was exhausted and I was working in London a lot doing late nights, which meant he needed to look after Fred. Then they both caught a terrible cold and spent a week (which felt like a month) under the duvet shivering, coughing and moaning every night. Lucky me!

In typical Charlie fashion, he allowed himself very little time for recuperation. The first day he began to feel well again he invited some friends to supper, but no sooner had he gone into the other room to blow his nose than he looked out of the window and saw the dog otter swimming by. He rushed through to the kitchen to tell us just in time. There was a stampede to the window and we all saw the otter get out at his old sprainting site and wander off into the trees. Charlie spent the rest of the evening justifying to our guests why he wasn't working and moaning about otters who turn up when you don't want them to. There is no pleasing some men. However, it was a blessing he hadn't rushed to the river that night because Fred took his first few steps, four to be precise. It was one evening when there could have been ten otters tap-dancing on the weir and we wouldn't have filmed or even noticed them.

The excitement and novelty of winter was beginning to wear off. The leaves were now completely off the trees and it was misty, damp and cold; the garden was bare and deserted and the temperature had dropped to minus three. Despite the tease of such a gentle autumn, winter was here to stay for month after cold month. We turned to the Bailey's and thought of Christmas.

It was around this time that one of the chickens died. She had been limping for a long time although we could find nothing wrong with her leg, and she was finding it difficult to get in and out of the hen house. We built her a ramp and then, for a while, kept her confined with food and water so that she could rest the leg. This seemed to work, and she did seem to be improving. However, her weakness had already set her apart and she was being badly bullied by the others. It was a side of my jolly, funny chickens that I didn't really want to see. They were evil. If she so much as walked too close to one of them she would get a good pecking. Somehow this gaggle of girls began to remind me of those horrible American college TV programmes in which any girl who is different is sneered at every time she walks past the lockers. It's not until she has implants and a nose job, dyes her hair blonde and pulls the local hunk that she is finally accepted.

I felt so sorry for our poorly chicken. The others wouldn't let her anywhere near their food any more so I fed her separately and tried to keep her apart from the others as much as possible. But she was losing weight fast. What kind of life was that for her? Chickens are not solitary animals, they like to be together. She became more and more miserable until she was so listless that she would barely move; the leg didn't improve and in the end she had to be put down.

Apparently it is in the nature of chickens to pick on the weakest in the group. They are not known for showing pity or offering sympathy and certainly not for changing their minds. I was sad to lose one of my brood so quickly, but she had never been right and, if I am honest, I was relieved not to have to witness her suffering any more. Once henpecked always henpecked, as Charlie will testify.

Just before Christmas the river froze. We had been hoping for snow as we were desperate to get some snowscapes into the film, but we had seen it only once in the whole time we had been here. However, despite the lack of snow, in the ice and frost it was winter wonderland time. The river was frozen from bank to bank. We filmed ducks all morning. We had no choice, they were so funny. There were seven in all, the most we have ever had outside the house at one time; they obviously knew where to come when the going got tough. On further inspection we realised that Trevor was among them. Trevor was the unusual black duckling we had looked out for the year before. It turned out now, though, that Trevor was a girl.

We got the kind of shots that we have all seen before yet which are perennially funny. As we fed them we watched them try to navigate the ice on their ridiculous feet, each orange snowshoe going in a different direction. Out of control, slipping, sliding, the ducks lost every ounce of dignity. And if you are a duck you can't even look at your feet to find out what's going on. Just when they had finally made it across the ice to a piece of bread, the feet would lose their grip, the duck would keep going and overshoot, and miss the morsel completely. The kingfisher watched all this hilarity from the bridge, hungry and miserable. You can't fish when the river is covered in ice and you can't turn to bread if you are a kingfisher. He sat beside the Christmas fairy lights that we had put out, looking for all the world like a Dickensian waif.

We were, of course, feeding all the birds we could feed. This is a tough time of year for the creatures that live in the wild, and I always feel desperately sorry for them. Every five minutes, blue tits, coal tits and chaffinches arrived on the feeders; we even had a pair of jays visiting that December. With all these birds, however, came trouble. For a few weeks a sparrowhawk assumed our garden was some kind of fast-food outlet. He would come by every day around mid-morning, defying the air with his speed, launching a surprise attack on any small birds that happened to be outside the kitchen window. One day I watched with my breath held as he chased a blue tit around the beech tree for what felt like minutes. My eyes could hardly follow the chase it was so fast. They whizzed up and down, to and fro, through the branches of the big beech as though it were a roller coaster ride. Finally the sparrowhawk gave up and the tit got away. Presumably, there is only so long a bird can hunt at that speed.

The week before Christmas, though, the sparrowhawk

struck lucky with his initial swoop. He caught a robin and perched on the fence on the other side of the river plucking and eating it opposite the house. Then he disappeared in a flurry of feathers and we haven't seen him since. I was sad to see a robin killed, particularly so close to Christmas, but the big freeze meant it was a lean time for everyone. Even the fox who lives in the top field was taking risks, appearing early in the morning in full view, desperate to find something to eat. It was a shame for him that he missed the two pheasants we discovered running around like a crazy doubles team in the tennis court. They were so thick that despite having found the door to get in, they couldn't find it again to get out.

We had hung the feeders in the trees above the weir, just far enough away from the bridge so that the seed shells didn't land all over it and attract rats. We had also, optimistically, thought that this position might deter squirrels. We hadn't reckoned on Stunt Squirrel. Once the ice had melted and the ducks had regained their dignity, we turned to Stunt Squirrel for our entertainment.

Stunt Squirrel was brave and fearless. He could climb anything and everything and hang over precipices; he was fit and agile and he never gave up. Each day, first thing in the morning and around halfway through the afternoon, he would launch his assault. He would watch the birds feed all day, and it seemed as though, while he watched, his brain was calculating the diminishing weight of the feeders as the food was taken out. When he struck he would soar, in a series of thrilling and heart-stopping leaps and trapeze moves, to the thin branches of the beech tree which thrust themselves out over the weir. Below him was a deadly drop. The icy water would carry his small body away in an instant; he would freeze or drown if the sharp rocks didn't get him first when he fell. Although his performance seemed comical, this nut-gathering strategy was high risk. Stunt Squirrel barely noticed the chasm beneath his paws; like a cat burglar hanging over a diamond in a case, he was entirely focused on the nuts in the squirrel-proof feeders.

'Squirrel-proof' is a great phrase, but what does it mean? It certainly doesn't mean squirrel-proof. During a series of experiments for a programme about clever animals filmed in 1999, we tested just about every squirrel-proof feeder on the market. They were all different shapes and sizes; some had huge climb-resistant structures strategically placed below the food, others were dependent on critical aperture size or shape to defend the nuts, others used perches just right for a small bird but far too brittle for the weight of a squirrel. We left them out for weeks, and try as I might I can't remember one that worked. The same producer, Mike Beynon, was the man who had filmed squirrels on challenge courses in the 1970s. Those shots, cut to the theme from *Mission Impossible*, made the squirrel famous as a clever, skilful, agile creature who could assess, compute and learn tasks quickly, much more quickly than any of us had previously given it credit for. However, the bottom line for anyone who likes to feed birds in their garden is that the squirrel is a nut-nicker, plain and simple. I preferred to think of our own garden as less a battleground and more an assault course. If the squirrel was clever and brave enough to complete the task, then he deserved the prize. He also deserved to be on *The Krypton Factor*.

So, twice a day, having watched and assessed how much the birds had eaten, Stunt Squirrel would make it to the furthest reaches of the beech tree and attempt to execute his plan. Instead of going below or level with the nuts, he would go way above them. He could tell the feeders were beyond his reach, so his sights were set on the strings from which they hung. He was too clever to bite the string. He knew if he did that he would lose everything in the frothing water below. Instead, he would get himself into position, hanging upside down in the tree by his back legs. This left his front paws free to grasp the top of the string and, hand over hand like some small furry pirate pulling in an anchor, he would try to pull up the feeder and its precious cargo.

Time after time he tried, and every time he did we thought he might succeed. Every time he tried the feeder would be a little lighter because the birds had eaten more of the contents and every time he tried we were terrified that he might slip. But he never did. He never got the nuts either, because the feeder was always too heavy and he would have to let go of the string before it dragged him off the branch to certain death.

We thought he deserved a reward and left a great pile of nuts for him on a rock by the path. He spent a happy afternoon trying to bury them in the frozen lawn and in my empty flower pots. I'm such a sucker for a good show that he got another pile on Christmas Day. We buy pounds of nuts every year and the nutcrackers lie there unused, so it seemed only right to give ours to someone who would really appreciate them.

The river froze again in the first week of January. This time it was less fun. Our water pipe runs underneath the bridge above the river – the only way it can get to the house from the road – and it too froze. We had no water for a week. The river came in handy, providing water for our buckets which meant that we could still flush the loo, and a very nice chap from the water company visited with lots of bottles of water which we could then refill at friends' and neighbours' houses. Although it was a pain, we managed. It is only when you are deprived of such basic necessities that you realise how much you take for granted.

We tried to seize the opportunity to get some footage of the otters on frozen water; we thought this would be at least twice as entertaining as the ducks. Jamie headed down to the carp ponds one night, as we had worked out they might be there, but there was no chance. Some kids turned up, drunk, and started smashing the ice on the pond. They made so much noise that there was no point waiting for the otters. There was no way they would come anywhere near the ponds that night. So Jamie packed up the equipment and quietly slipped away. The kids never even knew he was there.

After the November bonanza, we were beginning to lose track of the otters. Their whereabouts were becoming a mystery. We stared at the river until we imagined ripples. I was watching the icy water one night and was sure I saw otter ripples, but when I looked again I realised that the sheets of ice completely changed the way the water moved. I just couldn't be certain.

In the depths of our second winter at the house, we were beginning to recognise annual patterns in our non-human neighbours. A little early, yet possibly in anticipation of spring, the goldcrests were fighting. They are the smallest birds in Europe, smaller than a wren, with a wingspan of just nine centimetres, and look cute even when they fight. They are olive green in colour with little flashes of gold on their heads and white stripes on their wings. There are lots around the house and it is great to have them so close; you can watch them from the warm. Even at such close quarters, though, for most of the year they are difficult to spot. They are so small that the slightest leaf cover on the trees renders them practically invisible, so winter is the best time to study them when the trees are bare and dormant. They are very busy, constantly flitting about like nervous ping-pong balls, feeding on insects and spiders in the bark and having the occasional scrap.

The sparrowhawk was succeeded by a kestrel, who hung around over the paddock. Every day he caught something, hovering first for long minutes while he monitored his prey, and then swooping so quickly that the vole or mouse didn't have a chance. I can't look at a kestrel hovering without thinking of the old name for them. In the sixteenth century they were called wind fuckers, which suits them extremely well.

We put beech hedges all round the garden to keep out some of the wind and lend us a little privacy and spent the first week of the year building a luxury riverside residence for otters in an exclusive paddock perfectly located for easy commuting to carp ponds and the Avon. Well, I say 'we', but to be fair Jamie did the spadework.

I knew that otters have been known to use artificial holts, and so we built one. We had had the idea in the summer, and if it worked it could give us one of the most exciting shots we would ever get. If it didn't we would have wasted quite a bit of time and money, but it was one of those things that we would always regret not doing if we didn't at least try.

We knew that Bridge Holt, which our otters used regularly, was only a pipe which led to a single chamber, so that is what we replicated in the paddock. Pete from down the road went all the way to the other side of Bath in his truck for us to collect a huge clay drainage pipe which weighed a ton. We laid it so that it led from the bank into the paddock. Then Jamie dug a chamber approximately three feet by three feet, which we filled with straw and covered with branches and earth to give it the feeling of being underground. Jamie also smoothed the way from the pipe to the river, down the bank, to make it more inviting. We then placed spraint around the entrance to the holt so that it stank of otter, but it also stank of us and I suspected it would be a long time before they used it. After all, there was no shortage of places for them to lie up; they already had a great choice of locations.

So why go to all this trouble when we knew where some of the existing holts were anyway? Well, this was no ordinary artificial holt, this was an otter television studio. Inside was rigged a small infra-red camera, about six inches long, with a three-millimetre lens which gives a very wide shot, allowing you to see the whole of the chamber in one picture. All the cameras we were using to film the otters were sensitive to infra-red light, as it is outside their visible spectrum. So in theory you can light a location without the otters minding.

They can see a glow on the lamp but not the light it casts.

The camera had a lead which ran through the holt wall and outside into the paddock, where we could connect it with the minimum fuss and noise to a monitor and a recorder. It was wonderful to creep through the paddock first thing in the morning, full of trepidation, the frost clinging to the long tussocks of grass which remain from the last summer's growth. The tall trees follow the path of the river and very often rabbits leapt away from you along a network of paths they have created through the grass. You would crouch down next to the mound of earth, find the cables, connect them and the picture flicked up in front of you on the monitor so you could see inside the cosy holt. It is inviting enough to tempt you in. Each time you do it, you hope against hope that you will see them. We could all see the shots in our mind's eye: three otters curled around each other fast asleep.

Although we all accepted that it was unlikely the otters would move in for some time, we checked the holt every day. There were no signs of disturbance, not a pawprint. We had to be patient and make do with daydreams. If our female was pregnant she might give birth in our holt. Imagine how fantastic that would be. To watch her feed, caress and wash her babies, to see them open their eyes, to discover how long she left them for and when, what she brought them to eat, when they took their first swim ... But we didn't talk about these dreams, that might be tempting fate. It was extremely unlikely we would get such shots, but this whole project had sprung from some pretty unlikely events in the first place so we'd have been foolish not to try.

By halfway through January we had become very concerned at our lack of otters. We were worried that, having resided on our river for many months, the otters had had the best of the fishing and moved on never to return. Charlie and I also knew that our time with Jamie was running out. He was becoming popular at the BBC, and we only had six more weeks with him before he went to Kenya to film cave elephants. This made getting the otters on film even more urgent. So Jamie and Charlie donned their waders and searched every inch of the river. Up and down from source to sink they waded, and found them, eventually, just downriver from the house. They weren't in one of their normal spots but there was fresh spraint and plenty of it. And so the fun started again, or so we thought. That night, despite the copious spraint, there was no sign of any flesh-and-blood otter.

Charlie took the whole of the next day to walk up the river in his waders, this being the only guaranteed way we knew to confirm whether an otter is around. I wish I had known what he was planning. In the month after Christmas it would have been far better for my thighs to have gone up the river with him than to sit at my desk writing scripts. I also missed out on some exciting discoveries. It appeared things had been happening up at the manor.

There was a new holt. It was definitely an otter holt because there was spraint outside, although it was hard to tell exactly how old the spraint was because it had been raining. It was a hundred yards south of the manor garden, in a field full of cows, under a tree root in the riverbank about four feet above the water level. There were, in fact, two holes next to each other. Charlie looked inside but couldn't see where they ended. They might have been originally dug out by a rabbit or a fox, in which case there could be a whole network of tunnels and chambers.

So all we could do was make assumptions. We had no answers, and still no otter sightings for January.

A few days later, the sand spot at the end of the weir revealed that the dog otter had been past the house and up the river towards the manor during the night. We immediately put Jamie on duty for the following night while we drove to Manchester for me to present my regular Sunday morning programme *Heaven and Earth* for the BBC, and we kept our fingers crossed.

There was also one other thing. We had discovered that

our friend Colin, who did the garden, had seen an otter at lunchtime the previous Saturday where the river runs through the paddock. He told us he was going to put up a fence to stop people from wandering along the bank, as it seemed that a group of lads had been taking axes to some of the trees. The otters already had to contend with factories, pubs, roads and dog walkers. They would be unlikely to return to the river if they were continually disturbed.

So we knew they were around. The frustrating thing was that we weren't seeing them, and were certainly not coming close to filming them. And we still had a long way to go before we had anything like enough footage for the film. Jamie got nothing that night, but the otters were back a week later. Again we didn't see them, but we found some fish remains by the carp ponds where they had enjoyed a little light supper. We tried to remain calm. Time was slipping by, but at least they were still here.

Chapter Thirteen
WATER RETENTION

....

A level change

February began with floods, and on one Saturday morning the river came closer to breaking its banks than we were comfortable with and I nearly didn't get to Manchester for my weekend's work.

It had been raining hard all week. The river changes character completely at times like this, roaring and pushing, becoming a force to be reckoned with. This natural feature that most of the time runs through our garden spreading calm seems to suck the life from the earth around it. It is no exaggeration to say that if you fell in outside the house when the river is in this mood, then you would almost certainly die. The river powers over the weir and through the sluice with such force and speed that no one could survive. Whole trees bob about like mere corks. Logs float past, enough to keep us in fires for a whole winter. Then there is the noise, a constant rumble as though a thunder-cloud were trapped just outside the front door, which increases if the door is opened. If you stand on the patio you have to shout to make yourself heard.

It rained through the night and all morning, the weather gods obviously having no respect for Saturday sporting fixtures or the nation's entertainment. Not only was I due to travel to Manchester to prepare for Sunday's programme, which is live from 10.00 till 11.00 in the morning, but we also had Catrin and Mark staying with us who were appearing at the Theatre Royal in Bath that week as part of a tour of *Pirates of Penzance*. I had been to the performance the previous night and everyone had rushed into the theatre shaking umbrellas and catching their breath after running through the torrential rain. It was a novelty to go out for the evening and I laughed till I cried at Catrin swooning over the pirates. As the audience revelled in the melodramatic arias and rousing, complicated choruses, coats draped over the seats in the auditorium silently drip, drip, dripped onto the carpet until it was time to venture once again out into the teeming darkness.

Outside it was a similar story: drip, drip, drip from the trees, the bridges and the fields into the river, until almost imperceptibly its level started to rise. All those drips accu-mulated in one raging torrent meant there was every chance the river would burst its banks and stop swooning ladies, carousing pirates and TV presenters from leaving for work. By mid-morning we were beginning to worry. The rain was so heavy it was like fog; you could barely see through it and it was the colour of stainless steel. The river was one bubbling mass where the raindrops pounded it. The level was high at 9.00 and by 10.00 it had risen one and a half feet. We were constantly drawn to the window, our eyes fixed on the rising torrent. I craned my neck out and peered upriver. The paddock was starting to flood. It was time to open the sluice gate.

I am a modern woman. I am fiercely independent and firmly believe that men and women should be treated equally. I have learned, however, to be realistic, and sadly men and women don't always have equal physical strength, a source of endless frustration to me. I have an ongoing battle with that bloody rustbucket of a sluice gate, which seems determined to demonstrate my puny strength while I am determined to master it.

The sluice gate is in the middle of the bridge. It consists of two sheets of thick iron in a large frame which when raised allow water through, and when lowered restrict its flow. It was originally installed to allow the millworkers who lived in our house to control the flow of water into the mill, which is just downriver. Although now the mill no longer works and the river flows beside it, the sluice still needs to be used to moderate and control the river, and prevent flooding. I only battle with the sluice gate when I know no one else is in the vicinity. It is a private battle. The gate is opened and closed by means of a kind of corkscrew attached to the top of the frame. To open the gate you twist the corkscrew clockwise, this pulls the gate up; to close it, twist anticlockwise and the gate drops down. You move the corkscrew by inserting a special large metal pole horizontally into a hole at the top of the screw, then pushing on this pole until it forces the screw round. This is much harder than it sounds owing to the weight of the gates, the addition of large amounts of rust and the pressure of the water, and I have never been able to manage it. I always end up looking like a gnat impaled on the end of a pin, helplessly flailing my limbs in an attempt to gain some purchase, which is why I only do battle with the sluice when no one is looking.

This state of affairs bothers me as there is not always a nice strong bloke around when you need one (or a muscle-bound girl, before I am accused of sexism) and I can't bear the idea of being helpless.

Charlie was asleep in bed. We had last seen him the previous evening on the way back from the theatre. As we were driving up the lane two suspicious-looking figures huddled up in big jackets with dark fleece hats pulled down over their faces had crawled out of the hedgerow, clutching tripods and other oddly shaped equipment. It could only have been Charlie and Jamie. Not even burglars would be out in this weather. It would have been easy to pretend I didn't know them and drive on but I thought I should find out how they were getting on. So we stopped.

I wound down the window letting out a blast of warm air from the heater.

'Hi, how are you doing?' I called cheerily.

They grunted. Jamie looked in the other direction. Charlie glanced at him then back at me.

'Well, you know.'

'Nasty weather. Are you getting some good stuff?'

'OK, I suppose.'

Jamie looked at the ground.

'Where are the otters, then?'

'That's what we'd all like to know, especially Jamie. Come on, you arse.' And with that they trudged off up the lane, two hunched figures retreating down the beams from the headlights battered by the rain. I was tempted with the line, 'A cameraman's lot is not a happy one,' but on this occasion resisted.

They were extremely fed up, especially Jamie. He had just missed the shot of the month. Dedicated soul that he is, he had been sitting in the car night after night, for two weeks, waiting. We needed to illustrate how the otter traverses difficult parts of the river, and there is an industrial location a couple of miles down from the house near the village which poses a bit of a challenge to any creature trying to navigate its way upriver. By a factory the river is funnelled into a massive concrete bed, like a giant half-pipe, which incorporates a twenty-five-foot drop, concrete to concrete. Because of the funnelling effect, the flow here is shockingly powerful, even when the river is in a temperate mood. This restriction on the natural course of the river is not only because of its proximity to the factory but also to protect the village. In 1968 the river burst its banks disastrously and the pub was inundated by two feet of water. Immediate steps were taken to stop this happening again: the river was channelled through the deep concrete bed, which would hopefully contain it no matter how swollen it became.

There is no way an otter could swim against the enormous force of water here and so they have to avoid the whole thing. We had discovered the neat little route that they take. They leave the river, clamber up the bank and cross a road, go under a gate and through a field and then head down and back into the river. This was the shot that Jamie had missed. He and Charlie had spent an hour rigging two cameras so that they would get more of this journey on tape. The otter arrived but Jamie only saw it after it had passed through the first shot. He immediately pressed 'record' on the next one but it takes a few seconds for the tape to come up to speed and by then it was too late. His face had been very glum in the pool of orange cast by the streetlight but in typical fashion he didn't moan once. He had just been particularly quiet.

However, the important thing was that we knew the otter was on the river, so Charlie and Jamie had stayed up all night chasing him. But despite carefully calculating times and positions on the river, despite being rained on all night and lugging heavy equipment across fields in the darkness, they didn't get another glimpse. Charlie had crawled into bed at 7.30 a.m., bedraggled and depressed, with the prospect of a long drive to Manchester in front of him. So I was reluctant to reduce his sleep time.

But as we watched the river the next morning we could see it was getting dangerously close to the top of the bank. Only four inches separated it from the millpond and the front garden. We would have to open the sluice.

Catrin and Mark weren't due on stage till 2.00 p.m. but they decided to make a dash for it before the road became impassable. Living beside the river makes you relaxed about its moods, whereas guests tend to panic about rising water after a mere spattering of rain. This was brought home to me when no sooner had they made the decision than they were gone. I was left holding the baby, staring at the rusty sluice, the rain and the swirling brown water. Although accustomed to the regular rise and fall of the water, this time I was worried. Still, I thought, the show must go on; at least the audience in the Theatre Royal would get their money's worth. 'Gone,' said Fred, stating the obvious.

There was no way I would be able to manage the sluice gate on my own with Fred. One slip and he would be washed out to sea.

The phone rang. It was Catrin. 'Just to let you know,' her Welsh accent lilted down the line, 'the road is impassable. We had to go up the hill and through the village, and we only just made it.'

We would need to leave in just a couple of hours if we were going to make the script meeting in Manchester. If the road was impassable now what would it be like then? A Land Rover Discovery gave us a height advantage over most other drivers, but now something else occurred to me. Even if we could get out, how could we leave a house that was about to become a houseboat?

The water had risen another inch. There was nothing else for it. I wanted to wake Charlie calmly but I couldn't hide the rising panic in my voice, and my gentle rousing came out as, 'I'm sorry, but you're going to have to get up and do the sluice NOW!' I bundled the poor, dopey man into his long mac and shoved him out of the front door into the pouring rain, where the gravity of the situation dawned on him as he came to. The paddock was half underwater, which made it a water meadow, and the river was lapping at the steps to the patio. The large, flat-bottomed boat we use for filming was listing with the weight of the water she was carrying, and straining at the ropes which tied her to the patio railings. I prayed that the railings would hold.

When we bought the house there were no railings, and, lovely as it was to look down at the river, some kind of barrier was essential to stop young and drunk people from slipping, tripping or diving into the deceptively deep water. It had not been easy to find them. You can't put any old railings up by a beautiful river in such a breathtaking spot, so we had felt we should do the view justice and choose carefully. After a long search, I finally found just the things, hiding at the back of a salvage yard. They were beautiful. Well, they weren't beautiful at that stage but I could see that they would be. Taken from the inside of a cruise ship, they were wrought iron with peeling gold and white paint, and the design had a watery feel. Flowing curves and reed shapes took up the whole length, not one straight line in sight. I measured them with my breath held; we would need at least thirty metres to fence the length of the terrace outside the house, and that's a lot of railing. There was enough, just. They were perfect. My search over, I paid for them. They seemed like a bargain.

However, I soon discovered that in my haste to snap up these beautiful railings I had completely underestimated the amount of work required to do them up. First they had to be shot-blasted to remove all the old paint, then the sections needed to be welded together so that they appeared as one seamless length. This was also a much more difficult job than at first appeared because it turned out that not all the pieces were the same height or even had quite the same pattern. In my haste to measure the length, I had completely forgotten to measure the height. Luckily, our friend Ben had some free time and a welding kit and, for a small fee, spent quite a few days working it out and putting it together. Then, just when I thought the job was done, we had to find a way to erect them so that they would take the full weight of an adult if someone fell into them.

The River

Charlie called in Obelix, John Keenagan, an old friend of the family who looked exactly like Asterix's sidekick with his white vest, white hair, long sideboards that nearly met under his chin and a pair of lovely twinkly brown eyes. John turned up in his hand-painted turquoise van every morning, had a sausage sandwich made for him and left at 4.30 precisely. In between the sausage sandwich and home time he worked harder than any man a third of his age (and that gives you a clue as to how old he is). He spent days in the hot sun, drilling holes six inches in diameter and feet deep into the thick stone and concrete that makes up the terrace. It took for ever even with a huge drill, and at times it seemed the masonry was impenetrable, but finally the holes were done and the railings attached to posts sunk deep into the patio. I remember thinking that Charlie, John and Ben were getting a bit carried away with the safety side of things, and I questioned the need for such extra-ordinarily deep holes. The bill for restoring and fixing the railings ended up being more than the cost of the railings themselves. But now, as I watched the angry river, I knew that it was worth it. They held firm against the incredible strain. If Obelix had been there I would have hugged him.

Charlie started with the sluice, turning the stiff mechanism with what seemed like hardly any effort. It made me sick but had an immediate effect on the water level. Then he pumped the excess water out of the boat. This seemed to make little difference, she was still straining at the leash. The rain continued to come down in sheets. Still holding the baby, I gazed helplessly out of the window. But the water level was dropping measurably. Where I had only been able to see two inches of bank now I could see four. We could relax.

But how were they faring further down the river? We phoned our neighbours Dave and Marlene Pullen. They had lived on the river for at least two decades and so were used to her changing moods, but even they admitted that the water was uncomfortably high. Marlene confessed that it was all her fault for tempting fate: only the previous day she had had a new living-room carpet fitted. The water was washing at the threshold of their house. We wished them luck.

Now we were in no immediate danger, we needed to get filming. Charlie took a few shots by the weir and then jumped in the Land Rover – that was the only way out – and headed upriver to Poynton. We had got some shots of water rising there previously when we thought the river was about to flood, and now was the perfect opportunity to finish that sequence.

We had also erected some 'time-study' posts, set in concrete, at various points along the river. As they are absolutely level and static, you can set the camera up on these posts at intervals through the year and get exactly the same shot. The only differences are the seasonal changes – the amount or colour of leaves on the trees and the level of the river. For the finished programme you mix these shots together so that, in an instant, you can see how the river and its surroundings change from season to season. This technique can be used to stunning effect if the shot is well selected. The BBC programme *Living Britain* had used it brilliantly and both Charlie and I had been inspired to see if it might work as well in our film. This would be a perfect opportunity to mix through from the river flowing at a gentle pace in summer to the raging torrent that it now was, the type of sequence which clearly illustrates the range of conditions that the creatures living there need to deal with. The work which had gone into setting up those time-study posts could now pay off.

While Charlie shot off I turned my hand to packing; we would need to leave for Manchester pretty soon. Keeping a baby amused and out of all the drawers and cupboards he had no business in and packing for three people in thirty minutes kept me absorbed. The phone rang. It was Charlie.

'I can't find any of the time-study posts.'

'What do you mean?'

'They're all underwater.'

The river must be even higher than we had thought. I looked out of the window.

'I think you'd better come home, we nearly are, too.'

In the short time that I had been occupied with baby and packing, the river had had an apparent change of heart. Despite the open sluice gate it had turned again and risen quite dramatically. In fact it was higher now than it had been before the sluice was opened. This was particularly nerve-racking because we had no further means of controlling the thundering torrent pouring past our house. There could be no 'Oh well, if it all gets a bit scary we can open the sluice gate' – we'd already done that. On the opposite bank, the millpond was contained by only an inch.

I went into the kitchen and looked out of the window there. The small stream that normally trickled from the hill on the opposite side of the valley into the pool below the weir was now a river itself. Thick and brown with silt, it had swollen to five times its normal size and had already turned over a small reed bed. Washing across the newly seeded lawn, it had created a second weir where it met the river. The stream had completely burst its banks and was threatening to drown the little humpback bridge which crossed it. The next victim would be our drive. I held Fred close to me and suddenly felt very alone. I consoled myself by planning escape routes from the bathroom window.

Why live here? I asked myself. Normal people wouldn't. They wouldn't subject themselves to the whim of the elements in this way. Thoughts of London began to dance through my head: lazy Saturday mornings in Café Rouge drinking large cappuccinos, idly flicking through the papers and smiling as someone delivers a warm croissant. So what if it were raining outside? There was plenty to do – visit a museum, go out for lunch with friends, invite some people over to supper or to watch a film. Life was so much more civilised. There was none of this; you didn't have to wage war just to keep your carpets clean or worry about what you were going to wake up to. And there was always the Thames flood barrier.

The gates opened and the Land Rover swept into the drive, covered in mud. Charlie almost bounced across the bridge, grinning from ear to ear. Flushed with excitement and exertion, he checked the river and the sluice as he crossed. He flung open the door and the roar of the river came in with him. 'It's amazing!' he yelled over the din, pulling off his waders and setting his camera down. Water dripped from the end of his nose, at least I think it was water. It ran in rivulets down the contours of his mac and onto the wooden hall floor, which was already one big muddy puddle. He slammed the door shut. The noise abruptly decreased by several thousand decibels.

'Dada,' said Fred, once again contributing by stating the obvious.

'I've never seen the river like it. You should see it at Poynton! The whole road is flooded. It looks like a second river. I nearly had to pull one woman out. She only just got her car through it!' I realised that Charlie was having the time of his life. Not for him lazy mornings drinking cappuccino in cafés. Who would be there when you could be here, battling the elements?

'You really know you're alive,' he continued.

I pointed out that the river was still rising.

'I know. It's incredible. But it's stopped raining,' he said optimistically, ignoring the anxiety in my voice. 'Jamie's on his way.'

On cue the gates opened again and Jamie drove in. Charlie was outside like a shot with the kit, and they stood on the bridge exclaiming about the state of the river and deciding which would be the most effective shot.

I could see one shot for myself. I went to get my little digital camera and carefully set it up on the office window sill, pointing at the opposite bank. If the river was going over, I wanted that shot, the moment when the water began to pour over the front lawn. I pressed 'record' and waited. The water lapped seductively at the very edge of the bank.

I took a deep breath and picked up the phone. Several calls later I managed to track down our programme editor. The big boss was enjoying a weekend off. I was about to ruin it.

'Chris, I don't think I'm going to be able to make the script meeting. We're about to flood.'

'Don't worry about the script meeting, what about the programme?' His voice belied the calm manner with which he took the news.

'I'll definitely make the programme,' I heard myself saying, 'if I have to swim to Manchester. Don't worry, I won't let you down.'

I put the phone down and watched the little screen on the side of the camera. The water was still lapping but the clouds above were clearing.

Fred went to sleep.

Marlene rang. 'It's normally three hours after the rain stops when the river starts to drop,' she cheerfully informed me.

The boys were stomping around outside with a camera and various tripods. I looked at the little screen again. The water was right on the edge.

The next day, as from the comfort of our warm studio in Manchester we welcomed everyone to *Heaven and Earth*, Ross Kelly, my co-presenter, recounted how I had only just made it. The papers on the coffee table in front of us were full of pictures of abject riverside residents desperately trying to salvage their belongings.

Marlene had been right, the river started to drop rapidly almost exactly three hours after the last rain. We had got some beautiful footage of flooding from other parts of the river, which would create a dramatic sequence for the programme, but I never got my shot of the water escaping from its normal channel at the house, and for that I was really grateful.

We had been lucky. This time.

A lot of fresh silt had been washed up by the flood onto the steps by the patio, and the next morning I found mink footprints in it. We knew they hadn't been there the night before because the steps had been underwater, so they were fresh. Charlie positively identified them for me because my tracking skills, although getting better, are still a bit dodgy. We immediately assumed that the lack of otter activity had led to the return of the mink. There had been much talk in the area of mink numbers decreasing and we knew that this had coincided with the return of the otter. However, within half an hour we were proved wrong. You can never assume anything with animals.

In the paddock we found more footprints that indicated the otter had used a lie-up only ten feet upriver from the mink prints. It had clearly been used by otters for a while because it was well worn, and there was now fresh spraint. Despite the torrents of the night before, it had already been a busy morning on the river.

We checked the artificial holt in eager anticipation of a resident – even a mink would have made a great shot – but there was still no sign of disturbance. Nothing had even briefly been there.

But we were learning that with otters we had to be constantly vigilant, and just a few nights later that vigilance paid off. Suddenly, the otters were again putting in appearances all over the river and Jamie got his shot. Three weeks of sitting up in the car all night freezing to death had finally been worth it. We had the shots of the otter coming out of the river to avoid the concrete channel and drop at the factory. In typical loping otter fashion, he ran up the bank, checked for cars on the road, crossed it, and then we picked him up on the second camera running under the gate and taking absolutely no notice of the NO FISHING sign.

Later that same night, excitement reached fever pitch (it

doesn't take much to get us going any more). Charlie and Jamie had moved further upriver to try to catch the dog otter as he came through. By now we were rigging known sprainting sites so that we could get shots of the otters marking territory, but the boys had also taken a chance and rigged the outside of the artificial holt too. It worked. They got some good footage of the very interested female sniffing around the pipe that led into the holt. I would love to say that we then filmed her on the interior cameras as she went in, but she didn't. Patience was beginning to pay off and we weren't too disappointed. Her interest was registered.

It was rare to get so much in one night and we were happy. We were beginning to learn that for every good night there would typically be three weeks when nothing happened. During each spell of otter drought we would all secretly worry that they wouldn't come back, but each time they did we had to be waiting for them. The amount of work for the quantity of footage was daft, and when we calculated how much footage and how many individual shots it would take to make the film, we sobered up very quickly. Although we were having some success and this way of filming wild otters on Britain's rivers had never been attempted before over such a long period, we were conscious that, for a fifty-minute programme, we might not actually get enough.

Meanwhile, another dawn saw the otters go to bed and the rest of the world get up, and already there were signs that spring was looming. The grey wagtails had started to get territorial. Two males were chasing each other around, tweeting with irritation, flying high and then down across the water. It seemed early, but they were already carving out their territory for the mating season.

By mid-February Jamie had gone to film those elephants in the dark caves of Africa. In some respects, his work would be less difficult (an elephant is easier to spot than an otter, after all), but without him ours was going to be a lot harder. With the obstreperous wagtails giving a clear message that spring was coming, we decided to postpone the otter filming and concentrate on other creatures while he was gone.

The weather, however, had other ideas. More rain came. We were due for supper at the manor, which sounds very posh indeed but really is just the lot of us crowded round the Aga in the kitchen eating good food and gossiping. I had been in London for the day, where it had been fine and sunny, but the train stopped at Swindon and could go no further because the line was flooded. Luckily, I was saved by a kind couple who were going through the village next to ours on their way home and I gratefully accepted their offer of a lift. They dropped me off up the road from the manor. I got there just in time.

On this dark, cold, soaking wet evening, it was hard to tell what was going on at first, but there was lots of shouting and rushing around. It seemed that the water was pouring off the fields on its way to the river in such quantities that the drains couldn't cope, and half of them seemed to be blocked anyway. This was one flood too many; they had had enough. The driveway had turned into another small river and was lapping at the front door. On closer inspection I could see that water was beginning to get in underneath it. Until that time I had barely noticed the front door of the manor. I had certainly never used it and never seen anyone else do so. We only ever used the back door.

I exchanged stilettos for wellies several sizes too big, grabbed a spade and dug for all I was worth. All we could do was attempt to channel the water down the long drive. The rain kept coming, and the noise meant that we had to shout to be heard, but slowly our plan began to work. The water changed its mind about using the front door and began to head off to the river again.

Eventually, Charlie turned up. He had been busy with his own single-handed rescue down the lane. The river could no longer fit under the bridge and was pouring over the road. An extremely optimistic woman in an Astra had attempted to drive straight through it. The Astra was not so brave and had given up halfway. Luckily, Charlie had come round the corner a couple of minutes later and had been able to tow her out. He was soaked. She was apparently completely dry, but very grateful.

When we were all sure that the temporary drainage system was working and the front door was safe we retreated indoors. There we steamed gently in the kitchen and ate creamy fish pie. Smoked haddock and boiled eggs dissolved on our tongues and melted into the mashed potato topping, peas popped sweetly in our mouths and the wine tasted like the finest Chardonnay the world had to offer. For pudding, home-made profiteroles completed the transportation from the cold, wet elements to the comfort of the indoor world. The chocolate sauce was warm and smooth and the choux pastry light on the tongue but comforting and filling in the belly. A meal after any kind of outdoor exertion always lingers in my memory.

The extra calories came in useful when we got home. So cocooned had we been, we hadn't given another thought to the floods. Water covered the drive and was lapping at the garage. We had never seen the river there before and I was more than a little concerned. The garage is at least twenty years past its sell-by date so any more than a gentle lap would have it in ruins. Even I, the eternal optimist, had to admit that this was the worst flood yet, although there was something very exciting about seeing the house in such different circumstances, like under snow. But it is frightening, too, to know how vulnerable we are. We couldn't stand and stare for too long because we had to push our other cars and help get the neighbours' cars up the hill to safety and then get all the kit out of the garages in quick time. But the rain had eased off, and when we realised that the water had stopped rising we gave up. We collapsed into bed having survived the floods again. My Paul Smith suit, however, had not been so lucky.

★

Charlie spent the last week of February getting up at dawn to try and get footage of the kingfisher in a bleak midwinter setting. On Monday the bird flew straight past three times. On Tuesday he perched on top of the hide listening to Charlie's radio. On Wednesday the radio fell in the river. Finally, on Thursday, Charlie got the shot. He said it was one of the best he had ever got of a kingfisher. With all that work I felt it ought to be.

The artificial holt was still empty. We had no idea where our female otter was, or her cubs. We presumed that they had split up by now, which would be the natural run of things. It would be difficult for them to stay together now that the cubs were grown because together they consumed so much food. It was odd not to know what they were up to. All we could do was keep our fingers crossed that they were OK and would return at some point.

We were still hoping that the female was pregnant again. Every day we would look for signs of her. It would be fantastic to have her back on the river with new cubs, and we needed all the footage we could get, but we were well aware that she might choose a different river this time.

We had managed to get our hands on Final Cut Pro Three, a new product for our Mac computer which would enable us to edit at home. It wouldn't be the polished final version for broadcast, just a rough cut, but because we had the system at the house we could edit as we filmed. This was a huge benefit because we would be able to spot mistakes before we reached the end of our filming period. We would also know whether some of our more creative ideas had worked before paying a huge amount for a proper edit, only to find out that they hadn't.

This was vital to us mainly in terms of our own confidence. The long dark evenings were now spent wading through manuals inches thick, backwards and forwards, learning the systems and the short cuts. The fire would be lit and the red wine would be opened and then the swearing would commence, but finally we got there and cut our first piece to music. It was a flood sequence, way too long for the finished programme, but we were very happy with it and it proved something. If we carried on working hard we might just get a decent film out of all this.

Chapter Fourteen
TERROR FROM THE DEEP

....

A clucking disaster

We had discovered that there is a downside to keeping chickens, and that is the fox, the mink, the badger – in fact the whole army of predators just waiting to take a chicken off your hands. Although the enemy will occasionally strike in the day, most kidnappings occur after twilight and so, to keep them safe, we made sure the chickens were locked up in their hen house, warm and snug, by the time it was dark. For the first few nights, back in September, this had meant catching them, which in turn had meant a lot of squeezing through hedges and crawling under bushes. But they soon got the idea, and now they were quite happily putting themselves to bed. We just had to check they were in and shut the door.

It was all too easy to forget, though. There was one late afternoon, for example, when we were all happily sitting in the kitchen up at the manor. It was wonderfully warm from the Aga, Tess the golden retriever was snoring under the table, and we were indulging in that most wonderful English tradition, tea and cake. Stephanie politely enquired after our chickens and watched our faces fall as we looked at each other. Neither Charlie nor I had remembered to put them away. Even as we stared, dismayed, at each other over the cake on the kitchen table, the fox could be massacring them in the hen house, and it would all be our fault. Once you have that image in your head – all blood, gore and flying feathers – then it is difficult to carry on making polite conversation and eating cake, so we were home in about eight minutes flat. To our relief, the girls were happily chuntering away to each other on their perch in the hen house and the fox had not yet discovered this restaurant. They really were turning into very sensible chickens.

Except for another freezing winter's night. That evening, when I went to shut them up, only four were in. Edna, the one with the missing middle toe, was now missing altogether. I shone the torch around the hen house several times but there were only the four Flossies, all of whom looked at me in a pious, 'We would never do anything so naughty' kind of way, and none of whom could give me a decent answer when I asked where Edna was. I searched and searched the garden in the darkness but there was no sign. She might have found a cosy place to roost outside and wanted a bit of privacy, but I had to assume the worst: she had been got.

When she didn't turn up the following morning I gave up hope. It was a horrible feeling and I vowed to keep the chickens in their pen. No longer would I indulge them with the freedom to roam around the garden. Nice though it was, it was just too dangerous. People had tried to warn us that foxes will strike during the day, and it was my own fault because I hadn't listened.

At lunchtime, our neighbour Pete, who lives half a mile away, telephoned to inform us that he was frying an egg for his lunch.

'Nice!' we said, a little perplexed because he doesn't normally phone to tell us what he is having for lunch. But apparently this egg was special. It had been left under his flashy new motorbike in his garage by our hen, who was now sitting under his Land Rover.

We went to collect her, and found her covered in oil where she had tried to roost in Pete's old sump oil tray. In 1993, both Charlie and I had been in Shetland helping as volunteers after the tanker *Braer* ran aground, decanting nearly 85,000 tons of light crude oil into the sea. Charlie was on the beaches in the south collecting oiled birds and I was working in a rescue centre on the north of the islands for seals and mammals. We never met. However, this meant that we both knew that oil and birds don't mix very well, and that should she try to preen herself she would ingest enough oil to make her seriously ill, if not to kill her. We had no idea how long she had been sitting there.

Edna went straight in the kitchen sink, where we smothered her with washing-up liquid, which is what we knew you do with oiled seabirds. We tried and tried to wash the stinking stuff off with warm water, and Edna to her credit was as good as gold and just sat there quietly. Although this seemed like a good sign, it may have been because she was ill. We needed to be as swift as we could, but the more washing-up liquid we added, the more gunk there seemed to be. Gunk that would not come off. We were tempted to cut away the problem and take off the feathers that were really bad, but she would need them for insulation, particularly while she was recovering.

My accountant phoned. I asked if I could phone him back as I was in the middle of washing a chicken. It was only later that I wondered what he must have thought. Fred found the whole thing hysterical and wanted to get into the sink with the chicken.

We were a bit flummoxed. I began to scrape the oily brown gunk off her feathers with a fingernail, but to do her whole body would have taken hours and would still have left a film of oil for her to ingest. What we really needed was a safe agent which would break down the oil so that it could be washed off.

After a little thought, I phoned my old friend Les Stocker at St Tiggywinkle's. I had worked with him in Shetland and he had helped out at many oil spills since. It turned out that we had been right to use washing-up liquid but that there was only one brand that worked on oiled seabirds. Les paused at that point to explain that he had never tried it on chickens. It was Fairy Liquid – kind to your hands and also to chickens who have spent the night in a sump tray. He had one more piece of advice: 'Give her Pepto-Bismol in case she has swallowed any oil. Have you ever tried to give a chicken Pepto-Bismol?'

Charlie rushed off to our local shop to get the Fairy Liquid and medicine. While he was there, he picked up the latest gossip. Apparently, the previous afternoon, a strange chicken had been spotted here, there and everywhere. First of all, it had been seen sitting on the wall outside Judith's cottage, having a chat with a cat. Gill had chased a chicken out of the Old Mill kitchen; somebody else had seen it crossing the bridge over the river. Then it was spotted wandering down the lane. Finally, it gave Pete's wife the fright of her life as she was about to do the school run when it leaped out at her as she walked into her garage. The only person who kept chickens up that way was Marlene. Several people had phoned her but none of hers was missing. Charlie said nothing, which they must have thought odd because he is normally first with the gossip. He simply paid for the washing-up liquid and Pepto-Bismol, complained a little about indigestion, and ran home. It seemed that Edna had had an eventful twenty-four hours.

The treatment worked. After a long session in the sink and five minutes trying to get Pepto-Bismol down her throat, then a few nights next to the radiator in the laundry, Edna was fully recovered. It was hardly surprising that she hadn't wandered too far since. I thought back to my original dreams of keeping chickens and pictured the scene of blissful, rural domesticity pictured in the back of that magazine. My advice to anyone else thinking of taking up this hobby? Entertaining it may be, but relaxing? My arse!

The inevitable happened one Saturday morning as we were enjoying scrambled eggs on toast in the sunny kitchen. Billie Holiday was on the stereo, we were feeling very relaxed, the coffee was good, the river outside sparkled and all was well with the world. Suddenly Charlie sprang up from his seat and with one word swiped the French windows to one side and sprinted off up the garden. That one word, the F-word, made my heart sink and Fred fall off his high chair. By the time I had mopped up the tears and the egg, Charlie was returning down the garden path, making a throat-slitting gesture with his hand.

The word, of course, had been 'fox'. It was all over for the girls. While chewing his toast and gazing at the river, Charlie had glimpsed a creature with red fur and a bushy tail in the shallows under the trees. It had taken him a while to process this information and what it might mean, and the fox had known exactly where she was heading. So she got there before him.

After the great escape we had been keeping a close eye on the girls, especially Edna. We had just acquired a couple of bantam cocks from John in our local Spar. The cocks were brothers, and got on very well. The idea was that they would keep the girls in order. We meantime were being much more strict about keeping them in their pen. We had blocked off any holes with planks and ignored their pleas for freedom. They had plenty of room to roam, and that, for any chicken, should be enough.

The cockerels were completely ineffective. We called them Wally and Larry, and they seemed rather effeminate to us, prancing and preening, more worried about how their feathers lay than keeping the girls in order, but in the bird world it's the blokes that dress to impress. Moreover, they were not very good at doing the job. In fact quite the reverse, they were completely henpecked and whenever possible kept out of the girls' way. They had plumey tails with a green tint and reddish-brown bodies, big red wattles and showy, fluffy anklets to keep their feet warm. Beautiful as they were, our girls were not impressed in the slightest. Wally and Larry would sit on the gate and crow for all they were worth but the hens pretended they were deaf and stalked off in the opposite direction. It was as though they were offended by our decision, as if they were rebelling at the thought that they might not be able to look after themselves.

Wally and Larry were never allowed near the feeder if the girls were there, they were squashed at the crap end of the perch in the house at night and they were shown absolutely no respect at all. Which might explain why they reacted as they did when the fox came.

They hadn't been out long that morning. We had all gone down together and Fred had collected the eggs for breakfast. We had laughed at them and the bid for freedom they always made the moment anyone went near the gate, and left them happily pecking around for the corn we had scattered. The boys wasted no time in finding a high point and crowing from it. We meandered back to the kitchen for our lazy breakfast, and that was the last time I saw them.

Charlie rushed up to the pen to find four dead chickens and a fox escaping through the fence. She had knocked one of our planks out and was scarpering with one of the girls in her mouth. The others were clean. There was no sign of blood but no sign of life. Charlie picked them up and put them in a box. In just a few minutes all the girls had been killed and only one had been a meal. It seemed like such a waste. I only hoped it was a much-needed meal; at this time of year the fox might have cubs.

When Charlie came back to the kitchen we realised that there had been no sign of the boys. Were they dead too? Had they already been taken? It didn't seem possible in such a short space of time. We were stunned. I was very sad, but we had both known it was an occupational hazard of keeping chickens. Every single person we had spoken to had warned us and we had done what we could to protect them, but that hadn't protected us from the shock of losing them all in one fell swoop.

Did I say all? A loud, familiar crowing began outside, with just an undertone of hysteria. It seemed the boys were alive after all. Alive but still missing. It took us a good few minutes to locate the source of the row but we eventually worked it out and looked up. About twenty feet above our heads, perched on the boughs of a tall beech tree, were the boys. Now recovered from the shock, which had silenced them for a while, they were flapping and shouting 'Murder!' for all they were worth. Being bantams, or should I say lightweights, Wally and Larry had been able to fly to safety. Not for them the foolishness of risking their lives protecting a gaggle of girls who hadn't even registered their existence.

And so now they lived the life of Riley. They had all the perch they wanted at night and all the food they could eat during the day. There may not have been anyone to listen to their crowing but there was no one to ignore them either. Wally and Larry were very happy alone. But this carefree existence wouldn't last long.

Through February and March, the nights were not abundant with otters but there was one otter highlight: at long last we got the shot that had been eluding us for over a year. The otter went past the house, slipped out of the water and passed in front of the camera. We were recording. Even better, Charlie had been filming from the patio on the other camera. The otter had surfaced to have a look at him before continuing downriver and we had it all. It was a great moment, and it showed how close the otters would now come to the cameraman.

April was the start of the really mad months. Having focused on the nighttime otter filming last spring, we now found there was so much happening that neither of us knew quite where to start. The month was balmy, there was no sign of the traditional spring showers whatsoever. In fact it was so dry that we even had to water the beech hedge we had planted over the winter. I was badly distracted by the lovely weather and the lengthening days. That is the problem with working from home and suddenly discovering the joys of gardening. I was constantly battling with myself to stay focused.

Charlie spent a lot of time on the patio. He, too, was trying to work. He would sit on the terrace above the weir in the sun behind his camera as I sat in the kitchen hammering away at the computer trying to put some shape into the programme. Charlie's bit sounds idyllic, but I could feel the waves of frustration emanating from him, even in the kitchen. He was desperate to film grey wagtails making love (well, mating, to be scientific). They had already done it four times that morning and he had missed it every time.

Grey wagtails are extraordinarily difficult to film, the fairies of the river, flitting here and there, constantly bobbing and dancing, never still and never predictable. You can usually spot one, or often a pair, if you are near a weir or a waterfall. They are small, about the size of a long-tailed tit, but their swooping flight will often catch your eye. Their name does them no justice at all; it is too literal. They do have lots of grey on them and they do wag their tails, so they are easy to identify, but there is much more. As well as the grey they have yellow, and lots of it, a kind of citrus yellow that blends with various shades of buff and charcoal grey so that somehow they always look well dressed and smart, even when dabbling in slimy shallows looking for insects. Why not call them the yellow wagtail? Unfortunately for them, there is another wagtail with slightly more yellow who has a greater claim to that name. As for the wagtail part, these are no mere tail-waggers, they are dancers. They negotiate falling water and chase insects with agility and breathtaking grace. I suppose it is their way to survive. They cheep and chirrup constantly, all in a high-pitched whistle, and are always around, come deep winter or deep water. Even when the river is pounding over the weir, you can still see a grey wagtail out of the kitchen window.

Today the light was warm on the water and twinkling at the weir's edge. It was the kind of day when you notice flies, like semi-transparent ghosts, going about their fly business, and the kind of day when the flitting feathers of the wagtails, as they twist, turn, settle, bob and fly again, are at their most fairylike. But alas, to catch the fairies mating was too much to ask. Charlie spent that day and the next swearing on the patio before giving up and transferring his energy to something more rewarding.

There was plenty to choose from. We had been nurturing a plan to film nesting birds incubating their chicks since the previous year when we had come to appreciate how risky their position is. Every night brings a good chance that the nest will be raided for its eggs, and if the predator is an otter or a mink, it will not just go for the eggs but for the mother too. Duck or moorhen, either makes a tasty meal. To the casual observer sitting on eggs might seem the easy bit, but we had realised that it was probably just as terrifying as looking after chicks. Any bird on eggs, whether wagtail, duck or moorhen, is a sitting target. We wanted to convey some idea of the danger they were in and if any of the nests were raided this spring, we wanted to have shots of it.

The only way to do this without disturbing the birds was to rig up security cameras on a remote system, but we could not do this until they had selected the sites and built their nests. We decided to concentrate on the wagtail and the duck because they had both been raided the previous spring. Then we noticed that the moorhen had chosen to nest on the bank directly opposite the duck's nest, so we decided at the last minute to increase our chances by covering all three. It proved to be one of our better decisions.

The problem was, predictably, that the duck had decided to nest in one of the most inaccessible places along the whole bank. You might think, given the risks, that this was a sensible decision, but only in the warped logic of birds. We had filmed her, along with her partner, searching for a nest site up and down the river, and prayed she would settle on somewhere near the house. She had even tried out some trees up at the manor, which was very comical. Both went up one after the other, had a look at the view and judged the ascent and descent like a couple of elderly house-buyers. Ducks have been known to nest in trees, but only on the odd occasion. Our ducks, I felt, were odd enough to do so. However, they decided against the tree location and began their search again, further downriver outside the house. There they climbed into the half-barrel full of soil next to the sluice gate, which, during the summer, is full of plants.

It seemed a very exposed choice. The barrel was effectively on an island in the middle of the river. It was in full view of the house and right next to the bridge so people would be walking past with dogs and children. Nevertheless, just as many couples become enraptured with money pits of houses, this pair seemed to like our barrel, completely unsuitable as it was. We were delighted. She seemed very reluctant to get out of it and he swam around and around proudly announcing that they would like to make an offer. She even began to venture off in search of twigs to start her nest. We left them to it. But either they were gazumped or they simply got cold feet, because next time we looked they were gone.

We found them the next day up in the paddock, where they were much more settled, although their taste in accommodation had not improved. Our ducks had chosen an old compost heap of grass clippings perched precariously on tree roots and fallen branches. That part of the bank was gradually eroding as the river's course changed a little from year to year, so the location was not particularly secure. A survey would certainly have shown up some subsidence had they bothered to have one done. Worse, the neighbours were bad news, they were in fact, the egg-eating kind. The old cuttings heap was covered in holes, holes that rats lived in. Unperturbed, proudly proclaiming her new home with loud quacks, the female settled on her nest with every intention of laying some eggs. Again, we left her in peace, but it was going to be an interesting few weeks and we were almost certain of a nest-raiding sequence for our film.

We gave her a few days to settle in and to make sure she didn't change her mind again, and it was when we went to check on her that we discovered the moorhen had chosen to move in opposite. The riverside property market was on the up. The birds were like neighbours watching each other from behind their curtains. Each bird laid some eggs and we decided that we should rig. We would need to wait until each took a quick break from sitting on her nest.

After two days of waiting for the duck to get off her nest, just for a little while, we were becoming frustrated. We had crept about doing bits and bobs that wouldn't disturb her too much but there was a limit. Her choice of site was presenting us with real problems. We wanted to get a feel of how exposed she was sitting there, vulnerable from every direction, and to do that we needed to shoot her from every side. However, the clippings heap was so unstable, there was no way it could support the weight of a human being rigging a camera. The surrounding trees were our only hope but they were hard to get to from the bank because it was so steep. The only way we could get close to that nest was to build a scaffold tower in the water, climb up it and rig cameras on arms in the trees.

It was a long job and we had to do it in phases, short bursts while the birds were off their nests. They seemed to be particularly dedicated mothers, reluctant to leave the eggs for any length of time, which was great for their offspring but not for us. We would never have forgiven ourselves had they deserted. I know some wildlife film-makers would snort with laughter at how soft we were being over an ordinary duck, but that's the way we work. We like to make life difficult for ourselves.

As soon as the birds were off their nests we went silently into action. Because the only way we could get to the nest was by boat we had to assemble the scaffold tower from a rocking platform. It was not easy, mainly because the scaffolding we had acquired was bent and didn't fit together very well. It wasn't off the back of a lorry, just well used. There were several times when I thought Charlie might fall into the river and drown. As Chief Scaffold Bit Holder and Passer I got frequent filthy looks for my recurrent outbursts of giggles, which rendered me unable to lift the heavy pieces of scaffold more than a few inches. I could see Charlie's point of view. When you are hanging from a wobbly bit of hazel sapling with nothing beneath you bar a few nettles and a drop into a deep, cold river and supporting a half-built scaffold tower with your little finger, you probably need someone sensible, but after a few goes we finally got the platform up. I am only grateful we weren't inspected by BBC Health and Safety.

Jamie was back from the African caves. He has far bigger biceps than me and would have been much more useful, but he had better things to do and only bothered to turn up later when it got a lot easier. Once we had our tower built, rigging the kit itself was time-consuming and fiddly. In the end the whole operation took several days. Luckily, the compost heap was within cabling distance of the house. The kitchen now looked like the studio gallery for Channel Four's *Big Brother*, only a less glamorous avian version starring one moorhen, one duck and one wagtail. We were now on twenty-four-hour watch; their every move was being monitored and we just had to wait.

On 22 April, three weeks later, we saw our first ducklings of the year. They did not belong to our duck, who was still sitting on her nest, but swam down past her nest and through the paddock.

We were gazing out of the window eating breakfast when ten pompom balls came zigzagging down the river with their proud mother. We thought she had come from a nest in Five Acre Field. As they made their way through the paddock, our duck had obviously mentioned the excellent service we offer because they lingered. Mum was very hungry, and although I don't like giving ducks too much bread, I knew she wouldn't leave the water for grain because she wouldn't leave her ten pompoms undefended. I figured that today of all days she probably deserved a good hunk of white bread. She ate the lot as fast as she could, diving for the bits she missed first time round. The pompoms squeaked in reply to her gentle grunts and zigzagged around her, sometimes going far away but always returning. They were heart-meltingly gorgeous. Perhaps it is an evolutionary strategy that they live on the water because otherwise there would be the temptation to pick them up and crush them by accident with too much love.

Nevertheless, I nearly got the chance to do this later. The routine for the mother and babies established itself quite early: they would swim from the field downriver to our weir and back again, so they appeared at intervals throughout the day. It was impossible to work out how the ducklings had the energy to do this. They expended twice as much as they needed to because, like frenzied yachts in a race, they were forever tacking to and fro rather than going in a straight line. The journey was probably half a mile each way and by lunchtime they had done it five times already, which meant they must have covered at least fifteen miles.

Just after lunch we realised that the system didn't always work brilliantly. Cheeping loudly and doing very small zigzags close to the bank at the bottom of the weir was one abandoned duckling. He must have been the only one go over and it was quite possible that the mother hadn't noticed him go because she had meandered back up the river with the rest of the brood. He seemed unhurt but was obviously missing the rest of the family. We went to the rescue.

The poor ball of fluff was being tossed around in the turbulent water of the pool and was too small for either of us to grab, but in the end we managed to fish him out with a net. Charlie cupped him in his hands and he cheeped and cheeped but settled immediately into the warmth. Now to find the mother.

We half jogged up the riverbank and eventually found her right at the top of the field with the other ducklings. The only problem was that the hedge running alongside the bank was impenetrable, so we had no chance of getting the duckling to her. We would have to wait for her to come back downriver. Charlie took the duckling down to a little beach where the river was shallow and he could easily reach the water from the bank while I followed the family's progress up and down the river by listening for their banter. Finally they made their way down the river to the beach. After making sure the duckling had spotted his family, Charlie let him go on the surface of the water. Legs going round in circles, the duckling rushed over to his family, cheeping just as loudly as the others. He seemed to be perfectly OK. There were no signs of injury or shock and he was welcomed back into the fold as though he had never been gone. The truth is, I don't think the mother realised he had been away. We wandered back to the house through the meadow, smug at having done our good deed for the day.

It wouldn't be the last time we went to the rescue. Those ducklings needed constant care and attention, and every time we were tempted to leave them to their own devices we were beguiled by their cuteness and how the odds of survival were stacked against them.

The mother turned up again the next day but it was heart-wrenching to see her. During the night she had lost six babies. Now only four bumbled around her in the same happy way as when they had been ten. She was starving again so we fed her, but it was all very depressing. As dusk turned into dark night Charlie followed her with his camera, and we got a real sense of what she was going through every night. Despite the darkness Charlie's infra-red light meant that he could film everything.

After some hours she began to get agitated. She was so

accustomed to our presence by now that it had nothing to do with Charlie. There was something wrong. Charlie filmed her pushing the ducklings into the vegetation at the side of the river. As if they had been told what to do they stayed there out of sight and silent. She, meanwhile, began to quack loudly.

Suddenly Charlie saw the cause of her concern. In the back of the shot was the mink, slowly stalking his supper. The duck knew he was there but had very little idea of how close he was; Charlie was the only one who could see everything because of his light. The duck swam away from the bank and the ducklings quacking loudly all the time. Showing the unreserved courage of any mother no matter which species, she was acting as a decoy, drawing the mink away from her babies by letting him know exactly where she was. It worked. He changed direction and began to follow her, but he was getting extremely close. How far did the logic run? Did she realise that if he caught and killed her, then he would no doubt find the ducklings, too? Yet she had no other defence – no horns, no talons, no teeth – no other way of protecting her brood. It was fascinating behaviour to film, particularly because the mallard is so common that we all think we know them well. This nighttime sequence would show that they had another life that we, who feed them so happily in the park, know nothing about.

Just as the mink got close enough to lunge, the female saw where he was and launched herself out of the water in a clumsy take-off, flapping and calling his bluff. At that he gave up. Once the element of surprise was lost his game was up. He carried on upriver without a second glance at the reeds which held the precious babies.

After a few minutes the duck returned to her brood and allowed them back on the river, but she remained vigilant all night. She didn't – couldn't – once relax, and unlike in the daylight the ducklings were herded the whole time. They were not allowed to zigzag around wherever they pleased, she kept them close to the bank in a tight group and stayed on the outside at all times. The secret nightlife of the mother duck was turning out to be one of fear and constant watching.

The next day we watched the family dozing on the weir, sitting on the soft cushion of weed while the water flowed beneath them. The sun kept them warm and the ducklings napped with their heads under their wings like adults. Now and again one would fall over and then settle again, which gave away their inexperience a little, but it was good to see them switch off. For the mother duck, though, it was pretty much a twenty-four-hour shift.

At 11.00 in the evening, on 25 April, the duck we had been monitoring for weeks hatched her eggs. We were two days out in our estimation, we had thought it would be the 27th, but we were filming and watching just in case. Now we needed to know when they would be leaving the nest. This would be a critical shot for us and we had to be ready for it. We couldn't rely on the remote cameras alone because we would also need to film them as they swam away and because we weren't sure how they would get down from the nest, which was a good four feet above the water. Would they jump or clamber? Would they be able to swim straightaway or would they need a few trial runs first?

We didn't even know whether all the eggs had hatched or how many ducklings there were because the mother was obsessed with keeping them covered up. I had read that the chicks call to each other from inside the eggs to synchronise their hatching so that the mother isn't trying to sit on eggs and look after wandering ducklings at the same time. Whether this was true or not we were unsure. The only thing we did know was that our mother was certainly less comfortable on her nest. She had constantly to shift and adjust her position as the inquisitive chicks investigated their home.

In the kitchen we were in raptures, it was such an exciting thing to watch. Now and again a little head would look out from under the mother's wing and we would get a glimpse of downy feathers and miniature beak before it was roughly shoved under again by Mum. The rest of the night was spent flicking through every bird book we owned, desperately trying to work out when the ducklings would leave the nest. But we could find nothing in the books. We would just have to be ready for them, to sit it out and wait.

Across the river, on the opposite bank, the moorhen was still sitting on her eggs, and compared to what the duck was now going through, still in relative comfort.

Jamie, now back at work, offered to take the first watch while we got some sleep, but making this film had never been as straightforward as that. We had only been asleep for what felt like minutes when the phone rang. On duck watch, Jamie had seen huge ripples and had sneaked outside to look at the sand spot. Luckily, it hadn't rained and he could clearly see the tracks of our mystery guest. While Charlie got up he started shifting the kit.

By the time Charlie reached Jamie, our guest had raided the hospitality and departed, leaving another clue to his identity – the remains of a huge chub. Unfortunately, this was the chub that Charlie had spent the whole of our first summer at the house catching to show people. He had been thrown back into the river so many times he must have got used to being caught and released again. This time he wasn't so lucky. This fisherman was no amateur, and he was hungry.

The fish had most likely been caught under the bridge and dragged to the bank, which now looked like the site of a large-scale massacre. All that was left of the chub was the head and tail, a section of shoulder and a small pile of scales and bones. The tail was as wide as my hand, the jaw monstrous. This was the first time we had seen evidence of an otter in months and it was a good sign.

Jamie and Charlie made a swift dash for the manor to set up some cameras on the pond, then thought they might check the holt there only to find that the otter had been and gone. They may have been swift but he had been swifter. However, he would almost certainly come back down the river that night, so the good news was that we would have a very good chance of catching him on camera. The bad news was that, at a few hours old, the new ducklings would be a tempting treat for a passing otter.

I had been reading that the first two weeks of life are the hardest for a mallard to survive and that the number of young passing that critical stage is a key factor affecting autumn duck populations. Given the obstacles and terrors these tablespoon-sized bundles of fluff face, it is a miracle that any of them survive longer than a fortnight.

A dull day dawned, with the first rain for many weeks. It was warm and the garden drank up the rain as it fell, the soil like blotting paper. The young, budding, thrusting plants had drained it of all moisture. In the grey light we could tell that the ducklings were still in the nest; the unlucky chub must have filled the otter's belly. The ducklings waited until dawn was complete to dive from the nest into the water, and we wondered if this was a deliberate ploy to avoid some of the dangers of the night. Luckily, we had had time to get back from the otter tracking and get the kit ready again, although we were starting to run low on batteries, having had no time to charge them all.

The mother went first and called and called until the ducklings braved the drop into the water. There were, we confirmed, nine of them. After getting the shot we had waited for, an exhausted Charlie had some lunch and fell into bed. I was left on camera watch with three monitors trained on the river: one on the moorhen, one on a wide shot of the river and the third on a wide shot of the bank with the moorhen's nest in it.

It's a funny thing to sit in your kitchen watching a moorhen actually two hundred yards away sitting on a nest. I would find myself moving very slowly in case she caught sight of me and got a fright, and we often found ourselves talking quietly, even though we knew she couldn't hear us or see us. Fred was the only one who made no adjustment in the volume of his voice. The first thing he did every morning was to shout 'Duck!' and climb on a kitchen chair. When he had reached this lofty height he would grab the monitor, a tad too clumsily for his dad's liking, and take a good look at the duck or moorhen. Only when he was satisfied that they were still there could he get on with normal baby business around the rest of the kitchen. For the most part this involved taking everything out of the cupboards and leaving it on the floor.

With all these ducklings taking over the river and the moorhen still on her eggs, it was hard to remember that there were other couples reproducing too. It seemed that the wagtails were having less luck. We discovered they had abandoned their nest under the sluice. There had been three eggs in it at one stage and now there was only one. We suspected a jay. A pair of jays had been watching the nest very closely all week, and we had scared them off the bridge many times. We didn't get a shot of them actually raiding the nest, but the wagtails had disappeared. This was extremely unusual, and we wondered what could have happened to drive them away. After spending most of a day waiting for them to return Charlie collected the last egg and put it into the boiler cupboard just in case. Nothing hatched, however, and after a week we gave up. We had probably left it too long, and with no mother to keep it warm it had grown cold. To our relief, the wagtails soon returned and began building a new nest. We wondered if this happened to the poor things every year.

Meanwhile, the ducklings continued to flirt with danger on the weir. Having fished out the first one to fall we decided that this was something that should go in the film. It was almost a rite of passage for the ducklings, part of learning how to navigate the river before they could fly. As they got older we would watch them throw themselves off the top. We never saw one injured. It was as though, being so small, they just bounced off the sharp rocks. In a warped kind of a way we also enjoyed other people's reactions. They were always the same: shock and panic and calls for us to go to the rescue immediately.

It was a tricky shot to get and we had been trying on and off for days. Every time a brood came down the river we would dash out with the camera and wait. The difficulty was spotting which duckling was going to go first. It was like trying to guess which lump of two-pence pieces would be next to fall in an amusement arcade. They would all dabble right on the edge of the weir, eating the weed and algae that grew there. Every now and then one would go right up to the lip and, just when you were convinced he was the next to go, he would clamber back to safety. Charlie would always be following the one that didn't go while another slipped over the edge.

Success came on another rainy April morning. It was freezing cold and neither of us had a chance to grab a coat when the ducklings arrived. This time, as if they were teasing us, they really took their time moving up to the weir. It was the same old story: you would think that they were going and then they didn't. The rain pelted us, and as my shirt began to stick to my skin I pelted the mother with grain to convince her to hang around. Charlie cut a hole in a bin bag so that he could cover the camera but still film, and after half an hour in the cold rain just happened to be on the duck that went over. We got him falling right to the bottom of the waterfall. Charlie filmed nearly all of them as they plunged over the edge to join the growing group at the bottom and then, last of all, Mum as she flew down, gathered them together and led them off down the river, bouncing and bobbing through the grey, choppy water.

The more we got to know these birds the more we liked them. They had such stoic yet humorous personalities and they were such entertaining neighbours. And the best bit was when they came back up the river.

There was no way they could get up the waterfall the way they had come down, so they had to use the steps up from the pool to the bridge. After a few weeks this was a piece of cake, but at first the ducklings were so small that scaling these steps was a monumental task. There would always be much quacking and toing and froing from the mother. The smaller ones would get stuck at the bottom, leaping futilely and running backwards and forwards desperate for some way up. The mother never seemed to help much, she just quacked a lot. Meanwhile, those who had made it up would stand at the top making a racket. Eventually, they would come back down and join the others at the bottom, and they would all climb over each other, trying to get a leg up. It was exhausting to watch, and in the end I felt they had suffered enough. I built some mini duckling steps alongside the bigger ones with rocks and bricks. These were used at least five times a day and, I like to think, much appreciated.

Having conquered the steps, the cheeping troop would then proceed along the bridge till they got to the middle. There, pushing and shoving nervously, they dared each other to jump. Unlike the weir, the bridge had no rushing water to give them a little push. In the end, one would go for it, and once the first brave duckling had plopped in and surfaced then the others would follow. There was always one reluctant one left, and heart-stopping moments when the others began to swim away and you thought that he would never pluck up the courage to join them. But the prospect of being left alone eventually always seemed worse than the drop from the bridge and we would cheer as he overcame his fears.

The previous April the ducklings had been decimated at this time, and the mothers had been swimming around with only one, two or three ducklings. This year was not nearly so heartbreaking, with family groups of up to ten surviving, although precise numbers were problematic. On one occasion when our two families were hanging out by the house, one duckling left his family of four and joined the other group of nine. No one seemed to mind and he stayed.

We also noticed something neither of us had read about in

any of the books. The mothers tend to swim with their families up and down the same stretch of river all day, in our case up to the manor and down past the weir at our house. Sometimes when they turn up they are a duckling down, and then later they seem to have made up the numbers. Whenever they feel threatened, our ducklings make a beeline for the edge of the river. We suspect that if they get lost or frightened they simply wait, tucked in amongst the vegetation, for their mother and family to come back down the river and then skip out and rejoin the merry band. After our first few rescues, we realised that the best action was no action when ducklings were separated from their mothers. We should simply watch and wait. Mother nature has a way of taking care of her own and so do mother ducks.

Chapter Fifteen
SURFACE TENSION

....

*Things aren't always as easy
as they first appear*

By May the pressure was becoming difficult to cope with. Charlie and I were stressed all the time and argued constantly. With so much left to do we didn't know how to prioritise. Each task was weather-dependent and we were constantly reorganising to account for the unpredictability of the sun. Most tasks were also animal-dependent, and animals did things when they were ready and not to suit our schedule. We were never sure if we were doing the right thing at any given time, or whether we should be doing one of the other fifty-two things on the 'to do' list. The dream of working together was beginning to wear thin. Instead of teamwork and joking around in the sun there was bickering in the rain. It may have been spring but the weather was not helping either. We spent an unhappy morning catching up on our time-study posts. This should have been a lovely job – no waiting for animals, a whole morning pottering about on the river filming its beauty – but today we were dodging clouds and waiting for the sun.

This was not quite the idyll we had hoped for and everyday we lived with the knowledge that this was our last crack at each season, that if we missed anything now we wouldn't get it at all. Even when the sun came out, it isn't too melodramatic to say that fear chilled us. We had no experience and no past successes to fall back on. This was it, make or break time. All around us spring was exploding. Plants seemed to be growing inches by the hour, we couldn't keep up and the sun wasn't helping us out much in the lighting department.

But that evening our mood changed again. Our moorhen hatched three gawky black babies, around the same time of day as the duck hatched hers. Both parents were in attendance, the father having been very attentive throughout the period of incubation, taking turns on the nest and even bringing food to the mother. The new arrivals were fabulous, and unlike the ducklings very exposed in the nest so we could clearly see them emerging from the eggs and wandering around. We were stunned by how ugly they were. They were fluffy but their skinny wings stuck out from their sides at odd angles. Their feet and legs were huge, completely out of proportion to the rest of their bodies, but the best bit was that for some reason they each had a bald patch on top. The chicks looked like three miniature vultures clumsily waddling around and made us laugh as we sat in the kitchen watching their first few hours of life. We already had plenty of film and they hadn't even left the nest yet. It was just what we needed for our spring sequence and our morale: shots of chicks hatching from eggs.

The next day was a bank holiday and so for once it was easy to predict the weather. It would, of course, be miserable. Now that the moorhen chicks had hatched, Charlie was like a cat on a hot tin roof, nervous that he might miss them fledging. So, while other families spent the holiday together queuing at B & Q and relaxing, our bank holiday was no more than a token gesture. I had worked most of the bank holidays of my adult life but now, because we had a child and were a family, I had become attached to the idea that we should make the most of them. We compromised with a quick trip to the Bath Flower Festival with a friend of ours called Aiden. I was rushed around each stall and only allowed to linger long enough to buy some farm-fresh sausages. Aiden, seduced by the blooms, decided that he might want to start judging flower competitions, but before he could interrogate the judges we were ushered home after just an hour in case those three tiny chicks decided to take one step out of their nest.

As ever, though it pains me to say it, Charlie's instincts were quite right. Less than half an hour after we got home, we watched the first chick leaving the nest from the comfort of our kitchen as the bank holiday rain lashed against the window. Using the remote camera we followed as it navigated twigs and vegetation in its father's footsteps down the bank and then – plop – it was in the water. It didn't sink but at once swam confidently. The miniature vulture was off, with its long neck and the patch on the top of its head. On our black and white monitors that white patch really stood out.

Once we had seen the first chick safely into the water Charlie went off to the hide with the film camera so that he could catch the bits the remote cameras would miss and we were put on monitor duty to keep an eye on the nest. Aiden, Fred and I watched from the kitchen, eagerly waiting for the remaining two chicks to get their first experience of the water that was to become their home, a momentous moment in a moorhen's life. But they stayed with their mother and showed no interest in leaving the nest. Even when the father returned they didn't move.

Unbeknown to us gathered around the kitchen table, a tragedy was unfolding. What we couldn't see on our monitors was what Charlie was filming on the river. It seemed that the chick wasn't ready for his first outing after all. He swam for a little bit and then tried to follow his father out of the water but couldn't. Charlie filmed him, torn between letting nature run its course and helping, and then suddenly realised that the poor little thing had died. All of a sudden he just gave up, or maybe got too cold, but after a short struggle he became still.

When the father returned for the chick, it took him a while to work out that he was dead, and for long minutes he sat on the chick in the water making odd movements at bits of floating vegetation with his beak. It was awful to watch because it was as though he were trying to drown his offspring. It was only when we studied the video footage afterwards that we realised what his movements meant. His instincts were telling him that something was wrong and he was following his strongest parental instinct, trying to incubate the chick even as it died, trying to keep him warm and make a nest around himself and the little one. Once he understood that his efforts were in vain he moved back to the nest to help take care of the others, leaving the dead chick bobbing at the edge of the river.

The remaining chicks waited another two whole days before they left the nest. We didn't know what the normal timing was and tentatively assumed that our dead chick had simply left the nest too early, before he had the strength or stamina to cope with life in the water. Of all the threats that they face, it seemed so strange and so undramatic, such a waste, for a chick to die on its first trip out of the nest. Even after wondering how the ducklings in their first few days on the river could have the strength to swim against the current and the endurance to do it for so long, this event really brought home to us how these little beings are not just vulnerable to predators but to the sheer challenge of living.

The end of May was fast approaching and we were due to film the mayfly sequence. The mayfly is a key component in the food chain and ecosystem of the river and we had decided it would have a starring role in the film to reflect this. However, filming such a sequence was complicated and time-consuming so we decided to draft in an expert.

Richard Kirby is, among other things, a very good macro cameraman, which means he does the really close-up small stuff. This was something that Charlie had neither the time nor the talent (his words not mine) to manage. Richard had shot some beautiful sequences for the Natural History Unit so it was very exciting when he agreed to work with us.

Mayfly live all year round underwater as larvae, and are easy to distinguish from other larvae because they have two antennae-type prongs sticking out the front of their heads. They have a streamlined flattened shape so that the current pushes them down to the river bed; that way they don't waste energy trying to stay on the bottom. They live under rocks or in weeds and have slightly different habits depending on what kind of mayfly they are, but there is one thing that is universally important about mayfly and that is their role as fish food. They are eaten as larvae and particularly loved by trout, which hunt them at night, but trout also love eating them as adults, which is why fishermen make it their business to know everything there is to know about mayfly.

For all sorts of reasons it seemed that we would have to shoot the mayfly in a tank, but we had wanted to film everything in the wild and decided to try to do this as well if we had time. Our first task then, before Richard arrived for his first day's work was to build what is called a weir tank about thirty inches square by ten inches deep. This has a lower front to allow a constant overflow of water into a trough at the bottom which is then recycled. From the front of the tank you can thus film the surface of the water while you are level with it instead of slightly above it as you would be in a normal tank.

We ended up with just a few days to build the tank and fill it with water, rocks and mayfly larvae so that they would have a chance to settle in before we started filming them. At the end of May and in the first half of June, as any good trout fisherman will tell you, mayfly larvae struggle up to the surface of the water to live their lives in a day. When they get to the surface, they shed their skin in a moult and push through the surface tension which binds them to the river, they emerge in adult form into the day to herald the start of summer and boom time for any creature that likes eating flies. On the busiest trout streams in the south of England, such as the Test, this is quite a spectacle, as hundreds of thousands of mayfly break free from the river at the same time. After a few hours and another moult, they are fully fledged adults and free to find a mate. The males 'spin', a dance famous among fishermen, which involves showing off in front of the girls (not really surprising) in the hope of getting one of their many legs over. They mate on the wing. The females lay their eggs, dipping down onto the surface of the water to do so, and then, well, sadly, there is very little else left for them to do, and so they dance their last dance, weakened by so much activity during the day, and they die. It is a predator fest, the equivalent of holding a beer festival outside a stadium just as the FA Cup final is finishing.

At this time of year, every bird on the river and from the surrounding area is frantically stuffing the constantly gaping mouths of a nestful of chicks. Food resources are always limited so this mayfly bonanza is exploited to the full by wagtails, kingfishers, even magpies and, of course, those under the water too. Trout lurk just beneath the surface, waiting for any hapless and weakened fly in its death throes to skim the river from which it struggled such a short time ago. We are surprised every year by the variety of visitors to the feast. Nobody, it seems, turns down the offer of a free meal. Poor mayfly. It is Armageddon on the river and it would take ages to film it all.

At night the otters were hanging around again, although the weekend before the mayfly hatch proved extremely depressing. We had an otter with cubs and a mink on the river at the same time and lots of ducklings and moorhens and all we got was a single shot of a mink! Our one consolation was that it wasn't just any old shot, it was a particularly useful one: of a small female in the tunnel entrance to Bridge Holt. We already had great footage of the otters using that holt so it was nice to be able to show that it had more than one resident. The mink looked very stoat-like with her squashed-up muzzle and tiny tail. We knew a female otter was around because of the tracks, and she certainly had cubs, but we hadn't seen them and had no idea whether this was a new otter or the female from last year with new cubs. Either way, she had slipped back to the other, larger river into which ours flows. The mink was still around but we couldn't work out exactly where.

Each morning brought ducklings and their parents demanding bread and grain. Our ducklings were growing fast and were now well past the two-week danger zone. There were still nine of them despite the presence of mink and otters. They could manage the weir quite easily now, although they still made a lot of noise about it. Their dad had stayed with them, which, according to all the references I could find, was not the usual thing. This was a new-age father, a modern-day mallard. The ducklings were about half-sized by the middle of May, still quite fluffy but preparing to get their grown-up feathers. They hung out around the house a lot waiting for free grub, or sunbathed on the weir while the cool water rushed over their bums. It seemed a dangerous place to rest but they loved to have a nap there in the afternoon when the sun was in the right position. Some preferred to dabble rather than sleep but even Mum managed forty winks, her head tucked under her wing, only one side of her face visible and the eye closing despite itself. All you could see was the white circle of her eyelid in the middle of her brown feathers.

I spent a long time holed up in the kitchen reworking a schedule for the spring and the summer. Putting all the tasks and stress into little boxes and then fitting them together would make our lives a little easier till September, when we needed to stop filming altogether. It was difficult, but when I had finished life looked a lot better than it had done for a long while. Being super-organised, I had built in enough time for otter filming and made it all very flexible. There were even a few days off for Charlie, if I could persuade him to take them. But within two weeks it was all going wrong. By June it was apparent that this was to be the summer of no summer – great timing for our first film.

So far, we could count the number of balmy riverside days of golden sun on the fingers of one hand. For us this was disastrous. The gamble we had taken the year before, sacrificing daytime filming in favour of otter night action was beginning to seem like an awful mistake. It became impossible to stick to the schedule because the weather meant that any sequence we did try to film looked as though it had been shot in the middle of winter.

We had allowed two weeks to shoot all the mayfly footage. It needed to be that long because capturing any kind of predation on film means sitting and waiting. You either focus all your efforts on the victim-to-be, in this case the mayfly, and hope that your particular choice attracts also a trout or bird, or you find the predator, which is more tricky, and focus on that. What always seems to happen is that you start filming one mayfly and then, out of the corner of your eye, you see a chaffinch or wagtail about to nail another. So you swing the camera round as quickly as you can but just miss it. In the meantime a trout has eaten your original choice so you then swear a lot and start again. Before you know it an afternoon has passed and the sun is setting behind you. I have no experience of filming mayfly myself, you understand, these are merely the observations of a wildlife cameraman's missus.

And you really only have two weeks. It is generally over the last week of May and the first week of June that mayfly coincide their emergence from the water. Timing is crucial; the more of them that leave the water at the same time, the greater their chances of survival. This is a deliberate tactic to cope with the number of open mouths and snapping beaks waiting for them above the surface. It is known in scientific circles as predator swamping and works for mayfly in much the same way as herds do for other animals: safety in numbers.

So we were governed by the emergence of the mayfly and in no position to change our schedule according to our own needs. By the end of the two weeks we had very little footage and were depressed. It had been grey and rainy for the whole fortnight and Richard had spent all his time in the garage with the door closed. This is partly explained by the fact that we were using it as a studio. He had set up the tank and was waiting for a mayfly to hatch just inches in front of the camera. He is a very patient man. Lucky for me that he is; I kept forgetting he was in there and sometimes he had to wait hours not only for a mayfly but also for a cup of tea.

Charlie stomped around like a bear with a sore head and made lots of bacon sandwiches. Our budget was tight and we both knew that even if the mayfly condescended to extend their hatching period by a few days there was no way we could keep Richard on any longer. The mayfly binge is the difference between life and death for many of the birds. We could not leave it out, and underplaying it would be bad journalism. Our luck seemed to have run out this time.

Richard is a bit of a Charlie, in the sense that he has a dry sense of humour, and they have both been accused of being miserable old buggers. They are physically very different, though. Richard is shorter than Charlie, with a beard which he strokes often while he talks. He has a look of Grizzly Adams about him, but a tamed version. He speaks slowly and quietly and his eyes squint so you can't see inside. We talked about vegetable growing and mayfly, and indulged in the topic that arises whenever two or more BBC employees are gathered together: how budgets are being cut. I like Richard very much. He is gentle, and sitting in our dimly lit garage for a week waiting for minute mayfly to hatch out of a small tank with only the fridge-freezer for company can surely only lead to negative thoughts, so I can forgive him those. I have a suspicion he is slightly more cynical than Charlie, although I didn't dare voice that because I'd have hated to introduce any element of competition into the already shaky joke arena.

Standing under the green roll-up garage door with Richard, consuming tea and soggy bacon sandwiches while the drips of rain fell from the door to the concrete below, staring out at the bleak sight which was our garden and the river in June, was one of the most depressing moments of this film so far. It wasn't Richard's fault, it was the weather.

We tried to film anyway. We couldn't afford to miss the mayfly shots and we couldn't afford to pay Richard to stand around waiting for the sky to clear. Instead, we spent a whole afternoon with Charlie up to his waist in the river, trying to get the camera as close as he could to the surface of the water so that when a mayfly emerged, he could film it being eaten.

Ducklings love mayfly, and when they turned up halfway through the afternoon I bribed them with bread to go nearer and nearer to Charlie and the camera. They didn't mind Charlie at all; they were really tame now and associated us with good food. They particularly loved Fred because he had no sense of reserve when feeding them, chucking in huge handfuls of grain at a time.

They were now adolescents and surprisingly strong for their size, able to resist the pull of the water with ease, even when it was moving quickly. They happily performed for the camera. Bread can fill them up very quickly, especially if it is dry when they eat it, because it expands as it soaks up moisture. They say a kitten's stomach is the size of an acorn so a duckling's must be ... What? The size of a pea? Bread is therefore not the best thing to feed ducks but it was all I had, so I just gave them tiny little crumbs. I was worried they'd get full and paddle off to a more interesting location. Still, they seemed to enjoy the company and stuck around for the fun of it. Every now and again one would break away from the group, streak across the water and snap at a mayfly.

This event would be followed by a lot of growling and tutting from the man in the long rubber waders installed in the corner of the river, who had missed it again. The ducklings, needless to say, were far better at spotting the emerging mayfly than he was. But it wasn't a good time to take the mickey and, to be fair, the process is quicker than you might think. The mayfly swim up from the river bed to the surface of the water where they spend a while battling with the surface tension. It is as though they are trying to fight their way out from under a huge piece of cling film. This part of the process is difficult to spot from above, and once they break free they unfold their new wings and are away in a millisecond. It isn't in a mayfly's interest to hang around on the surface of the water, where they are vulnerable to predators from above and below.

We got the shot before sunset but we weren't very happy with it. It was too dark and there was no sense of the joy of summer. Even the mayfly, it seemed, were disappointed with the weather. It was hard to tell, but their numbers appeared to be falling already. Maybe their season was coming to a close, or maybe they had decided not to bother.

Friday dawned, our last day with Richard and our last chance at a good mayfly sequence. When I say dawned, I just mean that the grey sky got a bit lighter. We had come to expect nothing more. Lunchtime arrived: another bacon sandwich, Richard's last one, of the shoot. But then, at the eleventh hour, things suddenly took a turn. The clouds lifted, revealing the beauty of blue, summer skies.

To our joy, Friday afternoon went from grey skies and misery to heat and dancing sunlight in under an hour. We blessed our temperamental, unpredictable, changeable, wonderful climate. Just as we were about to give up it was all hands on deck. For all our planning and my scheduling, all we had was a few precious hours of summer. Charlie and Richard called Jamie in and they worked miracles. Together, they managed to get some fantastic emergence shots and some excellent footage of ducklings and trout going for the mayfly as they left the water. Our sequence was finished, just. It was also beautiful.

Chapter Sixteen
ALCYONE AND CEYX

....

Look it up; they were Greek

'A irplane,' said Fred as the tool bag flew past the kitchen window.

'Yes, love,' I replied. I knew it was a lie, but what can you do? It was so much simpler than explaining to a twenty-two-month-old boy that actually it was his father's tool bag, carefully weighted to mimic an Ariflex film camera, zipping down a steel cable from one rickety scaffold tower to another along the river. We needed a camera moving fast enough to film kingfishers. Sometimes, when you are twenty-two months old, life has to work on a need-to-know basis.

It was already July and we were nearing the end of our filming period. Only a month to go. I would dwell on just how frightening this was, but I realise this is becoming boring. What we were doing now was trying to get more of those spectacular shots, the kind that make programmes like *Blue Planet* or classics like *Life on Earth* stick in the memory, the kind that give the BBC's Natural History Unit its worldwide reputation for quality. The kind that win BAFTAs. And this particular shot, following a kingfisher at incredible speed, would show how the bird's skill in the air matched its skill in the water. It would give viewers some idea of the marvel that is this small blue bird.

However, although I believe in aiming high, there was every danger that we were getting too big for our boots. *Blue Planet* had had a budget of £5 million and was filmed over five years while our programme had had a budget of £200,000 (with fear of death or worse looming over us if we went over) and was being shot over eighteen months. Not that money equals creativity or happiness. It just means you can afford more swish kit and lots of people to help you. Still, look at what *Blue Peter* managed to do with all those toilet rolls on next to no budget.

The tool bag had been flying past the window for days now. It had taken six entire days to rig this cable and the dolly that carried the camera. First of all, Charlie and Jamie had built scaffold towers at either end. The first one, upriver from the house, rose as high as the trees. Perched up there, Charlie was level with the topmost branches. He could, apparently, see lots of dragonflies but got bored waiting for kingfishers to fly by. I know this because he was constantly phoning from the top of the tower to find out what was for lunch, or to remind me that there were some nice sausages in the freezer which might make a good sandwich.

Every time I looked at this construction it reminded me of the leaning tower of pizza (the scaffold tower, I mean, though this could also apply to the freezer), but when I tentatively brought up the subject of health and safety Charlie got quite offended. He informed me he had some qualification in scaffold-building which I didn't even know, or believe, existed. He then went on to express surprise that I could ever doubt his credentials and told me that not even I knew his hidden depths. I dropped it there. I didn't want to pursue his hidden depths if that's what they contained, so I carried on about my business which, at the time, was doing the washing. If all else failed, the cable dolly would make a spectacular washing line.

Gone were the days when I would jet off to Australia at a moment's notice. Now my world revolved around a very different set of circumstances and a lot more washing. Different, sometimes mundane, but certainly no less enjoyable.

After the towers came the cable. Using the boat it was relatively easy to lay it down the path along the river. The path followed the aerial track used by the kingfishers all summer. The closer we could get the cable to the kingfisher flight path the more likely we would be to get our shot. They came from the paddock and over the millpond, swooped low over the bridge, turned as they dived over the weir, above the weir pool and down through the tunnel of trees to the sandy banks. A little bit of clambering and some clever boat man-oeuvres and that was done. The cable was in place, but the real challenge was the speed. How would we ever get the dolly carrying the camera to go anywhere near quickly enough?

Visitors often have difficulty seeing the kingfishers because no sooner have we said 'Kingfisher!' than the bolt of blue is gone. You have less than a second to respond to the whistle which announces they are coming like some high-speed train round a bend. As you focus on them they are already level with you; by the time you say the word you can see the blue of their backs; by the time your expectant guests look up from the bottom of their glasses the halcyon bird is halfway to the sea.

If it were to take the weight of a camera and not droop – and to allow the dolly to move with anything like the speed required – the cable needed to be absolutely taut. So Charlie and Jamie had to tension it, using ratchets on six-foot-long pieces of orange strapping. This involved several more trips to B & Q, and another afternoon slipped by in a daze of sweating and cursing.

All this time the kingfishers were going crazy. The babies

were fledging, leaving their nests, learning about their new home and being introduced to the life they were about to lead. Mothers were chasing babies, tying to persuade them to be independent, babies were chasing mothers, trying to persuade them to revert to the good old days and feed them. Constant whistling and acrobatics accompanied Jamie and Charlie as they worked.

Usually when you yell 'Kingfisher!' in this house, everyone knows which way to look. They are nothing if not birds of habit, flying the same route almost to the millimetre. But now they seemed to be everywhere. It is the same every year; you know when the babies are out of the nest because there are kingfishers all over the place. It takes the new arrivals a little while to work out how to live in this multi-layered environment. They land on branches that swing wildly and are promptly bounced off again. Often we catch them flying far too high above the river, or losing track of it altogether because they have gone the wrong side of the trees and into the paddock. I saw one get so confused that he ended up following the wrong strip of grey and going over the gate and along the road.

It was not only great fun to watch but also heaven, as we were about to attempt one of the most challenging kingfisher shots that, to our knowledge, had ever been done. We were in the middle of a glut of kingfishers and would have no worries about subjects, even if we had some about which path they might take.

After the tool bag had been on its very last flight and the timing had been worked out to the second, we were ready to start. This would require three people: one to spot kingfishers, one to start the camera and release it from the top of the scaffold and one to catch it at the other end and switch it off. The third job was particularly important from a budget point of view. It was vital to switch the camera off as quickly as possible so that it didn't burn miles of expensive film. It would be running at ninety-eight frames per second, which is very fast but gives a fantastic slow-motion effect when you replay it slowly.

The problem was we could only afford two people. We were hoping to film at dawn because that is when the light is at its most magical; it can make the difference between a good shot and a spectacular shot. Somebody would have to look after Fred, and it isn't a great idea to have a baby monitor in the hide with you if you really want to get close to nature. So I was on baby duty.

After a ring-round, Charlie managed to find a guy who really wanted to get some experience working with a wildlife cameraman and was free that week. He didn't even mind getting up at 4.00 a.m. It was all going to plan. On the eve of the shoot, after a week of rigging and organising, Charlie went to bed early (well, straight after *Big Brother*) with all of the excitement of a child who knows he has a treat to wake up to.

But the gods had other plans.

The following morning, Thursday, saw the boys all gathered around the kitchen table, drinking tea and eating bacon butties – by 7.00 a.m. with glum expressions on their faces and a grey, dull, misty day outside. When it started raining they agreed to cut their losses and meet again at the same time the next day. After all, you couldn't really expect to be lucky on day one, they reassured each other.

Friday morning and the weather seemed fair. The sky was still white rather than blue, but the light was good and it was worth a try. They waited for hours but not one kingfisher made an appearance. It seemed the extra time spent rigging may have cost us much more than just an extra day's equipment hire.

Saturday was rubbish. No kingfishers and no blue sky.

Sunday we were in Manchester for *Heaven and Earth*. Reports reached us that the skies in the south-west were blue and we suspected there were kingfishers everywhere. When we came home we were greeted with clouds scudding slowly

over those same blue skies. We were also greeted with only
our second sighting of an eel. Charlie, miraculously, had never
seen an eel on this river before (the first sighting was mine),
even though the presence of otters indicated that they are
probably around because eels are one of their favourite foods.

We knew they were in the river for certain when Delia,
our next-door neighbour who had lived on the river for
twenty-six years, told us a bum-shivering story about her lazy
son's encounter with an eel. One long summer when he was
a teenager, he had taken to rowing a small boat out into the
middle of the millpond in front of the houses and spending
his afternoons being lulled to sleep in the hot sun by the
gentle rocking of the water while pretending that he was
improving his mind by reading an intellectual tome. This habit
ended abruptly one day when, thinking he was floating a little
too close to the shore, he took a half-hearted swipe at the
water with an oar in the vague hope that it would return him
to prime position at the centre of the pond, and inadvertently
flicked a huge eel into his lap, much to his shock and the
amusement of those on the patio. I'm not sure whether it was
David or the eel who hit the water first, but it is the prospect
of a close encounter with an eel rather than the ferocious cold
water that has persistently deterred me from swimming in our
otherwise inviting mill pool.

Then, in our first year here, I saw an eel as I was crossing
the bridge in the early evening. The sun, setting behind the
house and the hill, just reaches low enough under the trees to
light up the base of the weir so that you can see almost to the
bottom, and I saw one immediately below me. It must have
been at least two and a half feet long, although I am tempted
to claim three. It was green and you could clearly see its fins
sticking out at the side and along the top ridge of its tail.
Before I could call Charlie, who was only just down the path,
it saw my shadow and moved swiftly around the rocks until it
found a hole to hide in. I looked for it every time I crossed
the bridge for months afterwards but never saw it again. I

often wondered if the otters had enjoyed it; it was certainly enough to provide supper for our mother and her cubs.

However, this time Charlie saw the eel first. I don't think it was the same one; it was smaller, probably nearer two feet than three. (I might have been suffering from the size-distortion syndrome that afflicts fishermen and teenage boys, but I don't think so.) It was at the bottom of the weir. The pool was very low, owing to the fine, hot weekend we had just missed, and we weren't sure whether it had been forced out of some underwater residence or was attempting to migrate upriver. It was certainly looking for a way to get up through the rocks, swimming from one end of the weir to the other, raising its head to try to make it up the wall. We watched it long enough for Fred to learn to say 'eel' and for Charlie to say 'Actually I suppose I'd better film it' and dive indoors for his camera. Immediately he did so, Fred and I waved goodbye to the eel as it headed back down the river having given up its search. Charlie returned in a sweat, camera and tripod in hand, and we never saw it again.

The long-range forecast had warned of thunder, and on Monday Charlie took one look out of the window, called Jamie and told him not to bother coming. But Jamie turned up anyway a little later and they sat in the hide for hours underneath a grey sky with kingfishers turning over the weir to fly over the paddock instead of directly up the river. The scaffold tower seemed to be putting them off. Considering the interference they have tolerated from Charlie over the years this was completely unexpected. The thunder never arrived but the pressure was definitely beginning to build.

Tuesday. The weather was overcast again but hot and humid. And now there was one. Not only had we run out of free work-experience people but we had also run out of Jamie. This was a tragedy. We were going to have to sacrifice some-thing else in the budget to get him back. It would be impos-sible to finish this film with one cameraman and no assistant.

What we had left to sacrifice we weren't sure. Jamie, as an assistant, had already been getting a higher daily rate than either Charlie or I because we had squeezed what we were paying ourselves to satisfy the budget. So there was no room for manoeuvre there, and there were no luxuries to lose. We might even have to go back cap in hand and ask for more money.

Well, there was one last option: to go back to the pile of rejected assistants. There was only one – me. Charlie reluctantly agreed that it was my turn to help out. How nice, you might think, to sit in a hide on a boat on the river on a muggy afternoon, listening to the plop of fat carp and the buzz of insects, with no stress, just waiting for the paradise bird of England to make an appearance. I was very excited about my call to duty, and it would also give me a chance to relax. I hummed a little tune as I pulled on my waders.

'I've got one second to turn the camera on and release it, and it has to be halfway down three seconds before the bird comes through. OK? So it's really important that you let me know the bird is coming the instant it is past that elderflower tree. We're using channel one on the radios but keep it clear and only shout "Go!" if the bird looks like it is flying straight through. If not, then give me a running commentary, but make sure you do shout "Go!" when it starts coming through. Then the second you shout, I will have let the camera go down the dolly, so you have to get outside the hide, out of the boat, run through the paddock (where the grass is up to your shoulders), sprint round to the bridge and grab the string attached to the camera, pull the camera back up to the bridge and turn it off by flicking up the button in the bottom left-hand corner, which will be facing you. You need to do this as fast as possible so we don't waste film. OK? Right. I'm going up to the tower.'

With that he was gone and I was left on the flat-bottomed boat with the waves lapping against the bank. I flicked on the radio, channel one.

'Love.'

'Yes,' replied an irritable voice.

'Which one is the elderflower tree? Is that the second one or the first one?' Well, it was hard to tell in such shady conditions.

I sat down on my small box in my tiny camouflaged hide in trepidation. The insects were buzzing, gathering under the tunnel of trees. The lapping of the waves was slowing as Charlie's canoe receded. The paddock was quiet, as if before a storm, voles shrieked at such a high pitch that you could only just hear them, like tiny whistles from a miniature train very far away.

Sitting in a hide is not as much fun as you think it is going to be, even when you love spending time by the river with a passion. In the silence there is nothing to do but watch, and small things begin to get to you and scramble your brain. I could hardly see out. The hide manufacturers had been so diligent in shielding the presence of man from unsuspecting wildlife that they had hung several layers of net and muslin across the gap you looked through. I wondered whether they had ever tried spotting kingfishers. It is difficult enough without a criss-cross multi-layered muslin effect challenging your focus. I spent a while struggling to rearrange the netting with my head under two layers of muslin and behind one layer of large-weave netting. Slightly concerned that with its super-power X-ray vision the kingfisher might spot me, I kept trying to pull my face back from the net, but then the muslin would fall over my eyes again. It took a while, but eventually I reached a compromise that gave me some chance at least of seeing a tiny body hurtling down the river in the shade. The thought struck me that I had been here before.

The hours didn't fly past this time, either.

Other minor celebrities marry major celebrities and find themselves in a house in LA or on a paradise island, catapulted to the jet-set lifestyle. Not me, I find myself in a boat, uncomfortable and cramped, waiting for some animal to come along. You would have thought that night on the fishing boat in Skye would have been warning enough, but no. Caught up in the flouncing skirts of flirtation, I neglected to heed it.

A bee was buzzing insistently. I continued to stare through the netting, literally trying not to blink in case I missed a kingfisher. The bee was buzzing louder and louder. I kept staring. My brain started to go funny.

What if life worked on a need-to-know basis? I thought. So much information is pumped into our heads every day through eyes, ears, noses, jamming up our nerves and filling up our heads so we can hardly remember what we decided we were going to have for dinner. What if we only received the information we required? Ads on TV for example. You would only get adverts for new kitchens if you were thinking of buying one. TV programmes would no longer be able to bombard you with useless snippets, those 'Aston crawls' which trail along the bottom of the screen chucking useless bits of information at you like 'Jade's mum had cornflakes for breakfast before insulting the pope.' It would all go. Our heads would be clear, our minds free to indulge the creativity we all crave. How brilliant! I immediately got on the radio.

'Love, what would it be like if life worked on a need-to-know basis?'

'You don't need to know.'

I took that to mean shut up and watch for kingfishers, so I did.

The bee got louder. My stomach was empty and rumbled, but I could still hear the bee. That bee was really very loud. He was in the hide, I was sharing a small, hot space of around three feet by three feet with a big, hot bumble bee of around two feet by two feet. I waved my radio at him in a threatening manner. He landed on my waders; luckily they went right up to the tops of my legs. I tried to flick him with my radio aerial but he seemed to have superglue on his feet. I glanced out of the 'window' but couldn't see a thing because the muslin had fallen down again. The bee took off and landed on the box I was sitting on. I was just inches away from a very painful experience. Kingfisher or bee? Which one to watch? Was I a professional or a wuss?

I opened the hide and ran away. No, not really, but I did open the hide and shoo the bee out, and while doing so, I heard the loud whistle of a kingfisher flying past. This was bound to happen. Luckily, it was flying across the paddock and not along the river. Which was what they had been doing the whole time. The next day we watched them from the house. Every time a kingfisher approached it would go round the paddock, avoiding the cable dolly completely. We still had no idea why, but in order to get our shot and not end up having wasted the last week or so, we needed to work it out.

Do you remember me mentioning the hours I spent creating a schedule for the summer? What a complete waste of time that was. And I had been so sure it would be fantastically useful. You may not think a schedule is important. Let me tell you, in television, it is. It's the one thing that cements all those airy-fairy creative ideas into some sort of reality; it transforms the whims of our imaginations into a sophisticated to-do list and then jigsaws that into a timetable so that each hour, each day, another step of the creative journey is taken until you realise that, little by little, your vision has been filmed. That is how they are meant to work, anyway.

But, having worked in this business for some time, I have seen schedules crash and burn after the first hour. I have exchanged glances with cameramen in the early-morning light at the beginning of a day's filming when we have been handed, by an enthusiastic new director, a schedule which has no hope of being done in one day, even if we don't have lunch or breaks and don't get home until midnight. It is maddening when people allocate only thirty minutes to light a room, run through a long interview several times for different shots and close-ups, then allow no time to de-rig and move to the next location. I have always sworn I would never make the same mistake or create that much stress by drawing up an unrealistic schedule.

I had been proud of my summer schedule. It allowed loads of time for everything, because whatever you are doing – filming, decorating, making a cake or trying to get out of the door with a baby – it always takes a lot longer than you think. There was also, of course, the weather. By allowing lots of time for everything I felt I had insured against any bad weather. I had also built in plenty of what I labelled 'Otter Days' – days that could be swapped around to make up for the nights when the otters came and everything else had to be abandoned. Everything had been considered, which was why this masterpiece had taken so long to prepare. But even my schedule, loose as it was, had no chance of accommodating week after week of cloud. And however much he denies it, I am still convinced that Charlie kept forgetting to look at it.

According to the schedule, we were meant to have the kingfisher tracking shot done by halfway through July. We were now well into August, and it was still tipping down. The rain had been pretty constant since the end of April. Of course, the kingfishers, being completely egocentric, had not concerned themselves for a moment with our problems and had all but disappeared. Yet even after over a month of trying, the most fantastic husband-and-wife (well, barring the actual ceremony) team ever refused to give up. We did, however, have to abandon shooting outside the house because, after all the kingfishers had fledged and gone to lead grown-up lives elsewhere, there was nothing left to film. So we moved the whole thing up to the manor where, we had discovered, the kingfishers were slightly behind ours and the chicks were still in the nest. This meant we could rely on the parents coming and going up and down the same flight paths regularly with fish for their hungry brood. Thankfully, these birds didn't seem to mind our cable dolly at all.

We spent a number of days bored and frustrated, shouting at each other over the radios. I was in charge of looking out for the kingfisher, as before, posted in the small hide across the river from Charlie. He and the camera faced downriver, I faced upriver so I would be able to see them coming. But by the time I had seen one flash by, pressed the button on the radio and shouted 'Go!' and he had flicked the switch on the camera and let it go whizzing down the cable, the kingfisher was laughing in its nest and we were hissing at each other about who had the slowest reaction time.

These kingfishers were not having much luck fishing upstream and would be gone for forty-five minutes every time. Poor Charlie was in agony waiting 'like Jesus Christ on the bloody cross', which isn't as blasphemous as it sounds. He was standing up to his knees in water behind a sheet of army camouflage netting strung up between two trees. A couple of feet above his head was the camera, suspended from the cable by its dolly system that would carry it quickly and smoothly downriver behind the kingfisher. To maintain the speed of the camera, Charlie's end of the cable needed to be significantly higher than the other end. It isn't advisable to stand on a chair or box in a fast-flowing river, so Charlie needed to wait with his arms above his head holding the camera back, his fingers ready to flick the switch and start the film rolling.

He was very good. In the five hours we were there he only complained twice. But his arms went blue, except for the red spots from constantly being bitten by mosquitoes, and the back of his neck was bright red and swollen where he had brushed against some hogweed. It wasn't an attractive look, but it wasn't a good time to laugh. My legs went to sleep a few times and I got a nasty insect in my knickers at one point, but I thought it best not to mention that either. Each time we missed the bird returning to the nest, Charlie was livid. Another forty-five minutes of crucifixion to endure!

As I was staring at the river, determined not to miss the slightest blue flash, a small, grey, fluffy shape meandered out from the reeds, cheeping and thrusting its head backwards and forwards. I thought perhaps the wonders of being in the countryside might help take Charlie's mind off the agony of filming. 'I can see a baby moorhen. It can't be more than a few weeks old. It's on its own and very cute,' I told him over the radio.

The voice that came back over the radio didn't seem full of the joys of nature at all. 'Well, if you had been watching out for kingfishers instead, you might have seen the one that just arrived at the nest.'

I was furious. How dare he talk to me as though I were an amateur!

'It didn't come past me!' I snapped back. Charlie was obviously not convinced. I think his private opinion of girls in hides is that they are liable to be thinking of interior-design issues and looking at their nails. However, he is far too clever to voice such thoughts.

The truth is that I don't do my nails when I am given such an important job to do – there is very little point when you have loads of equipment to lug back to the car. On the other hand, you can combat interior-design issues, like whether the kitchen should be blue or green, while you are watching a river. And I only looked twice to check a grass snake wasn't creeping up behind me, and even then I kept one eye on the water.

After a kingfisher arrived outside the nest as if by magic a few more times, we realised that one husband-and-wife team was trying to outwit the other. They had seen us coming and were flying over the top of the hide, round the back of the trees and dodging in just behind Charlie's head to get to their nest. We moved my hide a few times to try and confuse them and give me a better view, but it was no use. We needed a plan, and Charlie had one. Anything to get this shot.

Bright and early the next morning, keen to get started, we opened the curtains, took one look at the rain making circles all over the river and clambered back into bed to enjoy a nice cup of tea, some bread and jam and a cosy family breakfast.

Finally, the rain shifted and it began to brighten a little. We loaded up the car and wriggled into our waders, still wet from the day before, stinking, cold and uncomfortable. With barely a word, we arrived at the river and rigged as fast as possible. We didn't really need to speak. We both knew what to do and how quickly we had to do it. Occasionally, if one of the kingfishers approached, we would sit low on the ground until it had gone so that we didn't disturb them. They had tolerated us up until now and showed no signs of being bothered by our presence but we weren't taking any chances. The raindrops lingering on the trees showered us as we bent branches out of the way to fix the cable dolly into place. The sun was out but not very warm, and this part of the river was still dappled with shade. To get the best shot we would ideally want clear light with no shade at all, but that would mean waiting a couple of hours for the sun to move round. In that time, with the unpredictable weather we had been having recently, the sun might well have gone altogether. We would have to compromise. It wasn't worth the risk of waiting.

It took only about fifteen minutes to rig in between king-fisher flights. This time my hide was in a different place. A little further upriver from where I had been the day before was a well-stocked minnow pool. Charlie knew it well. It was just behind a bend in the river where the curvature of the bank protected it, and the calm water meant that the minnows were easy to spot. Yet every day the kingfishers flew past it, apparently because it lacked a convenient perch. To judge their dive took long minutes of watching and calculating, and that required a perch. We had seen kingfishers hovering earlier in the summer, so we knew they could fish without a perch if they really had to. Charlie, incredulous, had filmed it, the bright blue bird against a rare blue sky, looking more like some exotic humming bird than a native British species. Almost level with the top of a telegraph pole, it had hovered like a kestrel, staring down into the water. That was the first time in all his years of watching and filming kingfishers on this river that Charlie had ever seen one do this and so we knew it was rare behaviour, presumably because it wasn't worth the amount of energy expended per fish. Therefore, we reckoned that if we provided a perch by the minnow pool, they would take advantage of it.

We positioned a beautifully smooth, slightly curved, deliciously springy branch so that it stuck out from the bank. I sat just three yards away, inside my hide, on a damp tree stump with my feet in the river. There were goosebumps all the way up my arms and legs and most of the time I shivered. My little net window offered the perfect view of the perch. Flies buzzed around me, but I clutched my radio as though my life depended on it. Which, to be honest, it probably did.

We were expecting it to take a few hours for the kingfishers to use the perch, although we had no doubt that they would notice it straight away. Despite their speed, they seem to see every detail as they fly up and down the river. It was actually only a few minutes before the kingfisher alighted silently outside my window. My heart leaped into my mouth. I radioed Charlie.

'She's on the perch.'

'Tell me when she leaves.'

It seemed an unbearable length of time. The kingfisher bobbed and looked about, then peered into the water for a while. She watched the hide, and although I was behind the net curtain I was sure that, with her special polarising eyes, she could see me. When she chose to ignore me, I didn't know whether to feel relieved or offended. Then, at last, a dive. The bird disappeared from the bottom of my small, square field of view but then, just as quickly, returned to the perch, but with no fish. More bobbing. A turn of the head to judge the angle and then another dive, this time successful. One large minnow had its head thwacked against the branch and was then turned delicately in the bird's beak to head for the stomach.

My finger trembled on the radio button. I had expected

the kingfisher to fly off with the fish to feed it to the chicks but then, I suppose, every parent has to feed itself too, and this parent showed no signs of worrying about her brood. She caught and ate another fish. She dived again. This time, for sure, she would head downriver. No, she ate that one too. The sun was fully out now; conditions were perfect for the shot. The kingfisher settled down and did a little preening. Perhaps she liked this new perch a little too much. In the end I was sick of the sight of her. She stayed for fifteen minutes, bobbing and settling on her perch in the sun, having a well-deserved rest. Finally, she showed signs of taking off, but did so in the wrong direction. I heard her land on top of the hide. Always expect the unexpected, I said to myself between gritted teeth.

I radioed Charlie and told him in a whisper, 'She's on top of my head, on the hide.'

'You have to watch out for her leaving then.' Easy to do, of course, when the kingfisher is sitting on your head. However, she wasn't comfortable there and, like an arrow, she shot past the window and down into the water, then back to her perch with a fish. A few quick tosses to get the fish in the right position and I realised that she had no intention of eating this one herself. This time it was head first. She was on her way back to the nest. The moment she took off, I radioed Charlie.

He rolled the camera along the cable on its dolly for a few feet to get it moving and then let go. After that there was nothing he could do. I sprinted down the riverbank as fast as I could in my waders and just managed to catch the camera at the other end before it swung into the trees, and switched it off. We had it, Charlie was sure. He had watched the camera tracking the kingfisher all the way. However, to be on the safe side, we did it again. More waiting and more aching, but we knew that when it came to the edit, we would have a choice of shots.

In the end the first shot was so lovely that we used it at the very beginning of the film. It showed the kingfisher's view of the river, with the halcyon bird flying slow-motion in the foreground and then entering her nest.

Charlie's neck was scarlet from the hogweed for nearly a week, and it swelled up horribly. He didn't mind. It really had been worth all the pain and time.

Chapter Seventeen
POOLS AND RIFFLES
....

Depths in the shallows

As our fortunes changed, so did the weather. Summer decided that she was ready to make her entrance and we were able to get some of the shots we had been waiting weeks for: the trout plopping into the river, some transition shots from above water to below and a glorious sunset landscape of the river valley. With the blue sky and warmth, my dreamlike visions of what it would be like making this film were close to coming true.

One August afternoon I spent in waders, shorts and a bikini top filming in the river. It was one of those days where you don't want to be anywhere else, a true halcyon day. Even more remarkable, I was actually allowed behind the camera, and not just to stand about and make useless observations, but actually to operate the precious Ari!

We had, all along, had the vague idea of using Charlie in various obscure shots through the film so that the viewer got some sense of who the narrator was, but we didn't want his story to take over. We certainly didn't want the film to feel like it had a presenter. The natural inhabitants of the river were always the stars. Now we thought that since we were into the last few weeks of filming we really ought to clarify those ideas and shoot the rest of the material we needed.

We needed some shots of Charlie fly-fishing. A fisherman is to trout what a rain cloud is to humans, something that obscures the blue sky and threatens to ruin all the fun, but fishing is one of the reasons why people enjoy rivers so much. But then who would operate the camera? We had long since run out of money for an assistant. Charlie would have to reveal his secrets and teach me how to do it. At last I could take control.

We decided to position ourselves below the weir in front of the house. Here the sun caught the water in a magical way, the perfect setting for the sparkly shot we wanted. So, early one afternoon we clambered down the slimy, moss-covered steps with all the kit. The sound of the water cascading over the weir filled our heads. Despite the time of year, because of the amount of rain we had had over the last few months, it was as loud as ever.

It's very dark down there. The huge beech and sycamore trees spread their branches over the top of the pool as if they are trying to hide it. If you look up you can barely see daylight. Each leaf takes up its carefully allotted place in the sun so that it can drink in the valuable light and manufacture as much food as possible, and virtually no gap is left. High above the weir, the house gleams white in the sunlight which you can no longer feel. I shivered in my bikini top and began to regret wearing it. I'm typically British, I suppose. One sunny day and I get overexcited. Only minutes later I was really regretting wearing it when I realised how many biting insects were sheltering from the heat of the day in this cool spot. Carrying lots of heavy, expensive kit, it was impossible to slap them off.

We waded downriver towards glowing patches where the sun penetrated the trees and cast the most beautiful dancing golden light. Well, Charlie waded, in his manly manner, striding in long green rubber waders with a tripod slung casually over one shoulder (Do you know how much those things weigh?) and a film camera on the other. I slithered and slipped about ten feet behind him, moaning, carrying very little and swearing lots.

The rocks under the weir pool are like a mini mountain range. Imagine a whole series of rocks carved by the water into triangular shapes with the sharpest point sticking up, and that is the floor of the weir pool. Tucked underneath each triangle is a cave whose residents, as well as trout, include eels.

Something I have neglected to mention about myself is my phobia of snakes. I have had this phobia since I was small and there is very little I can do about it. My nightmares all feature snakes and, yes, I know the phallic jokes. Snakes render me helpless with fear. I hyperventilate, my heart beats so fast that it makes my head swim and I feel as though I am cemented to the spot with no control over my limbs. It doesn't take much more than the thought of snakes sometimes for this to happen, and it isn't just snakes. Anything that moves in a snake-like way has the same effect – long worms, for example, or eels. I have tried to overcome this fear – it is a horrible thing to experience and hardly useful if you like exotic holidays and filming wildlife -- but have spent many nights in Africa and Australia too scared to sleep because of the mere possibility a snake might be near.

On the whole I keep quiet about my phobia because I find it embarrassing. However, I will confess that in the summer I do now stamp up and down our own garden path because Charlie once found a grass snake on it. I reason that if I stamp then the vibrations will let them know I'm coming and they will slither off before I see them. I don't know if I have a tendency towards this sort of thing but after I first saw the film *Piranha* I would sieve my bathwater as it came out of the tap just in case there were any fish in it. After I saw *Jaws* I was unable to swim, even in a pool, for months.

In an attempt to control this snake thing I even volunteered to have Paul McKenna hypnotise me for one of his television shows. It certainly worked. In under an hour I was happily holding a snake, exclaiming how wonderful it was and how beautiful the patterns on its skin were. But after a few months the effects of the hypnosis wore off, and the next time I encountered a snake, in the reptile house at the zoo, the fear took control again and I had to get out quick.

Now that I have confessed my phobia, you will understand the significance to me of these caves being a home for eels. All it took was one brush of a stick against my wader and my automatic phobia system took over. Helplessly, I watched Charlie's back receding as he strode further and further down the river. One foot was on one rock but my other was wedged under a different rock just inside one of the caves, and I could not move. My body would not obey orders, it felt as though the rational part of my brain was shrinking fast. I couldn't speak either; in fact I could barely stand up. My breathing got faster and faster. I hung onto the two boxes of kit I was carrying, terrified I would drop them into the water. Meanwhile, the really wicked part of my brain was creating images of eels, disturbed from sleep in their cave, writhing over my foot and making their way up my wader. I know – it sounds pathetic. I knew that I was standing in the middle of a perfectly peaceful river in rural England and that even if an eel got into my wader there was very little it could do to me anyway, but I was paralysed with fear. It coursed icily through my veins and turned my muscles to stone.

At long last Charlie turned round. He smiled – I must have looked ridiculous. I couldn't yell or move; all I could do was look at him with pleading in my eyes. He finally understood and, with the sunlight behind him, like some mythical knight he set his kit safely down on the bank and came back to rescue me. Owing to my numb legs, which wouldn't walk, he had to carry me. When at last I was sitting on a rock, further downriver where the water was shallow and nothing nasty could be lurking, I was still shaking so much that I couldn't stand.

We didn't say much. I apologised for being so weedy and he agreed, and then we got on with our work. Determined that my first shot on film as a proper wildlife cameraperson would not be shaky, I sorted myself out while he set up the equipment.

In due course we were ready. I still had the shivers but felt much better. We needed to move quickly because we were losing the light. The shot was a simple one. Charlie's feet would enter frame in the shallow, sparkling water (perfect in this location). I would track the camera along with them for a couple of seconds and then slow it down so that his feet moved out of frame, leaving only glittering ripples and dancing sparkles.

We practised a couple of times and then, just as we were all set to go there was a problem. I noticed, on the left of the frame, a stick waving about in the shot. I looked up from the viewfinder, but couldn't see it on the bank. Charlie went over and cut away a bit of dead vegetation with his machete and got back into position. I peered through the viewfinder and there it was again. 'Hold on!' The sun was dropping fast and Charlie was fast becoming infuriated, but I couldn't understand where this stick was and we couldn't film with it there because it looked ridiculous waving about.

Charlie marked his position with a rock and waded all the way back to where I was. He looked at me. I looked back at him, smug in the knowledge that I was doing the right thing. After one glance through the viewfinder he was helpless with laughter. Apparently, there is an exposure meter inside the viewfinder which waves about as the light changes. It looks just like a stick. Had he told me that in the first place, I wouldn't have held up the filming. As it was, my shot turned out to be a little masterpiece, and we used it as the penultimate image of the film. If I am honest, though, I don't think I'm cut out for the job of camera assistant. And although Charlie is far too nice (or perhaps sensible) ever to say so, I rather think that as an intrepid camera person I am probably more trouble than I am worth.

Chapter Eighteen
THE SEDIMENT SETTLES

....

And all becomes clear

Whenever I take the train back from London I revel in the changes of scenery and somehow seem to know that I have chosen the right place for my home. The light changes as you travel west, the trees stand tall against the setting sun, the spaces become wider and the sky gets bigger and they swell my heart. As the sun shows me the way home I pass hay bales that some bored farmer has sometimes stacked diagonally to amuse himself. I often spot a fox running across a field on some business of its own. Occasionally it will look up at the passing train in disgust.

Now the sun glints on the underside of a kestrel's wings, now green grass from an early-mown field is saturated with colour in this most flattering light. Mobiles ring up and down the carriage. A man snores so loudly that I can hear him three rows back and with the woman next to me I burst out laughing. A man up the front drones on into his black plastic box about how some poor guy called Stuart probably isn't up to the job, but outside . . . Outside, a hare tears across a field of golden stubble, racing the train, powerful and solid, thrusting forward with all his might, and yet we leave him behind. I see the White Horse high up on a hill on our left, flattered by the contrast with the mud of a newly ploughed field in the foreground. Wisps of cotton wool decorate the blue sky while a vague half-moon appears to remind us that this beautiful day is almost at a close. Black and white Friesian cows graze around the burned-out wreck of a once blue, now black Triumph Herald, and the train bounces so that our drinks nearly come out of their cups. Rosebay willow herb draws the eye with patches of startling purple. Rivers wind their way through fields rarely seen, certainly not noted, by the businessmen who are too busy assessing each other, but I know, and I feel privileged to know, that each holds a thousand stories.

As if I have been allowed to see enough, we slow down and enter a tunnel of trees. It's hard to know just how much more beauty your brain can take. Tonight I think I appreciated it so much because of where I had spent the day. I had been in St Thomas' Hospital, and there, while I was waiting to do a piece to camera, had been unable to prevent myself watching a man have his chest cavity beaten into breathing by a machine. A man who, because of his condition, looked like something out of *Star Trek* – pale, deathly pale, with big eyes and a big nose. He reminded me of the guy from the anti-smoking campaign on TV when I was small. He couldn't breathe for himself, so a machine strapped around his body pummelled the living daylights out of him like a small washing machine on a spin cycle.

Outside the ward was another river. The Thames sparkled in the sunlight and the world enjoyed the blue sky. Opposite, in the Houses of Parliament, men and women discussed the future of our country, but this particular patient knew none of it, and cared not at all. A complete stranger helped him empty his lungs while a machine encouraged his heart and body to work. Why, I wasn't quite sure; he had reached the end of his life.

This can happen to you when you are filming medical stories. The appalling nature of what we humans sometimes have to go through just to survive shocks me into being especially grateful.

A black, thick-furred cat disappears into the thin strands of green grass on a patch of wasteland.

But today it frightened me so much that for a moment I didn't want to go out into the world any more. Another man was wheeled into the ward prostrate on a trolley.

We pass geese, a whole field of geese, bright white in the sun.

Four years ago he dived into a pool and has not been able to move below the neck since. He communicates by sucking and blowing through a straw. It could be any one of us.

A mum, dad and baby watch the train pass and wave.

The magic light is going now, and the buffet car is closed. The sun has got too low, it can no longer cast its light so far, the shadows are longer than the bright spots. A meeting has been arranged to assess Stuart's potential for the job. Tractors, one blue, one red, are parked at the edges of fields. You notice these things when you are the mother of a boy of almost two.

A fox with cubs, quite old, almost as big as her, stands in the middle of a field of black and white cows. A rare treat. Gardens and caravans.

Brakes again. The sun gives one last gasp, throwing the last of its rays. A plume of smoke from a bonfire, familiar hills . . .

Maybe it's just the large dollops of red wine sloshed into plastic tumblers between the train's lunges, but I do love the journey home.

The last weeks of filming were like wading through treacle. They had that slow-motion quality that gives a sense of the unreal. We could not comprehend that so much time had actually passed, even though we always knew that it would, and now our window of opportunity was closing. Now the time was nearly gone, we should know how the film ended and feel as though we had finished shooting it. Instead, we felt as though we should be going to Mike Gunton's office and asking if we could start again now that we had a better idea of what we were doing.

All the way through, as early as when we first had the idea, one of the things we both felt very strongly about was keeping the storyline as real as possible. This might seem obvious but often, particularly in natural history film-making, the stories are scripted first and then shot to give the best idea of how the animal behaves or of what might happen to it in the wild. So, a wild dog in Africa will probably come up against a gang of poachers in his lifetime and that will probably be a dramatic or dangerous encounter. But no matter how dedicated your cameraman, it is highly unlikely that you will stumble across this encounter and then manage to film it in a beautiful way. You probably wouldn't even be able to get the lens cap off in time.

With these limitations in mind, and anticipating the requirements of the viewer, that is to say a programme that shows the trials and tribulations of the natural world, you can begin to understand why most programmes are scripted or storyboarded first and then shot second. And although this makes perfect sense, neither Charlie nor I wanted to make our film in that way. The joy of the river that had inspired us to make the film in the first place was that life on and around it was unpredictable; we were following certain characters, our neighbours, and from day to day we did not know what was going to happen to them. We felt we could give an honest account of life on the river because we lived beside it and so would always be there when something happened. We had the privilege of being beside our characters all the time, day and night, and could simply grab the camera whenever we needed to film an event in their lives. In the early days, getting the shot of the mink taking the kingfisher convinced us that this method was right. This film then was a great opportunity to illustrate life and the natural world as it truly exists on a river, not filtered by the way we wanted it to exist, and we were now looking for ways each character's story would resolve itself.

One day the end of one story came floating down the river. Underneath the avenue of trees upriver from the house floated the body of a kingfisher. We had always presumed that when they died kingfishers just got eaten, and that is probably the case for most of them. Charlie had certainly never seen a dead one in the river before. But there it was, and although it was immensely sad, it was at the same time one of the most beautiful things I have ever seen.

The head of the tiny bird was turned to one side and both wings were spread out, almost as though it were flying across the surface of the water. The trees that framed the picture filtered the bright sunshine of this August day and gave it a cool, green haze. Green leaves that had dropped into the river surrounded the dead body as though they were petals scattered by mourners. Most beautiful of all were the shafts of sunlight that penetrated the green canopy and picked out the intense colours on the halcyon bird's back and wings, massaging his feathers as the water moved him through each dapple.

We filmed him from the boat, cutting the engine and just drifting along beside him in the current. It was so slow and gentle. Any stronger and it would have sent the small bird spinning, but the speed seemed somehow graceful and appropriate, part of the river's laziness at this time of year.

The kingfisher seemed to be a young bird and might have been killed in a territorial dispute, but there wasn't a mark to spoil his perfection. Maybe he was one of those unfortunate souls who, having left the nest, never really get the hang of fishing and eventually starve to death. At first it seems incredible that this might happen, but when you have watched a kingfisher at work for any length of time you wonder that any of them ever manage to master the skill at all. Only a quarter of fledglings survive to breed the following year, a terribly sad statistic.

★

The biggest achievement of the last week of filming was to film the opening sequence. This had remained unchanged in our imaginations since the beginning. We wanted an image that would show how long Charlie has been in love with this river and at the same time evoke everybody's memories of playing by a river, which is probably the closest many people come to getting to know one.

We decided that the image should be two boys fishing with nets and jam jars. Because it was the very beginning of the film it had to be as beautiful as possible – golden light, sparkling water, a real sense of nostalgia – and for the most beautiful light we would need to shoot it at dawn or dusk. Some friends agreed to lend us two of their sons, so Joss aged nine and Will, six, and their dad Andy came to stay the night in readiness for an early start. The night before, they ate burgers and sausages from the barbecue, watched films and played martial arts with our coal-scuttle tongs. At 10.30 they were still far too excited to go to bed so they ate home-made ice cream and watched just one more video (while leaping off the sofa in Superman fashion), and by midnight we were pinning them down under their duvets.

Despite the late night, our film stars were up bright and cheerful at dawn, nets in hand, eager to get started on the fishing. Both are dark with long dark eyelashes and big brown eyes, round faces and winning smiles. Will is the more focused and slightly more serious. Joss flits from one interest to another, fascinated by everything but for less time than his younger brother, who will sit and think about it for longer. It was a cloudy morning, which was all the more frustrating given that the whole of the previous day, August bank holiday, had been perfect filming weather. There was nothing for it but to have breakfast.

After a few hours hanging around waiting for the sun to come out our leading men were getting restive. They had picked apples. They had learned what sloes were and picked some of those too. They had played a little tennis and fished with the new nets. But all they really wanted to do was embark on their new acting careers.

By lunchtime there had been a few glimpses of sun but none of the magical golden light of high summer that was essential for the beginning of our film. So we had lunch. Tins of spaghetti, home-made bread and more ice cream, which was the only thing keeping them going. Andy and Charlie told stories of how they had become friends in Shetland, which were only really interesting if you were Andy and Charlie, and the boys were now bored. We would have to wait some more and pray for a beautiful evening. Fortunately, and not for the first time, someone up there was listening.

Charlie and I had spent the afternoon getting some final shots at the quarry. The river passes through it, and so do the otters. It was dusty, hot work. We were filming diggers and loaders and graders, those big machines that tip stones from one to the other and then onto conveyors – every little boy's dreamland. As the mother of a little boy I can testify that there is something genetic which makes little boys compulsive about diggers.

The sun grew stronger and stronger through the afternoon and by 3.30 Charlie and I were exhausted and covered in dust, ready for a shower and a cup of tea, but the sun had burnt all the cloud away. This was our chance, our one and only chance. Between 5.00 and 7.00 looked as though it would be perfect. We dashed home and grabbed the boys and the kit. As we started to set up we knew we would be fighting time. The light would only be right for a short period and we had a lot to shoot.

We headed for the water meadows up at the farm beyond the manor. The farm is also a riding stables and the horses spend summer afternoons wading through the water to cool their ankles and flicking flies away with their tails. The river meanders across the meadows with rolling banks, gentle and shallow. At the near end of the meadows is a lovely, wide set of falls and then a bridge under which the river flows towards the manor. It is a particularly wonderful place to be on a summer's afternoon and idyllic for boys fishing for minnows.

There are always unforeseen problems when you are on a shoot and generally more when you are in a rush. We didn't know we would have to spend ten minutes pulling up the long grass beside the river so that it didn't tangle up in the dolly wheels and bring our lovely, smooth tracking shot of the boys running along the opposite bank to an abrupt and shuddering close. We hadn't bargained for the horses who lived in the field being so obsessed with getting their bums in shot, or being quite so reluctant to move. And we hadn't anticipated that huge stars who have been kept waiting all day might, under the pressure of too much excitement, throw a wobbly.

Charlie was poised on the small dolly, only just managing to keep his balance. I was switching between the roles of producer, director and camera assistant, and had one of the hardest jobs in the business to cope with: pushing the dolly, complete with camera, tripod and cameraman. Now you might think this sounds easy, straightforward manual labour, but think again. The whole thing is quite heavy, so it is difficult to get going, and once you gain momentum, even harder to stop before it rolls off the tracks. The aim is to maintain the same speed from beginning to end, which is especially tricky on a short track. We had had a few practice runs during which I had in no uncertain terms been told to concentrate on one job at time – to look at the wheels rather than where the boys were. In other words, to be an assistant rather than a director. This was understandable since Charlie, the dolly, the tripod and the extremely valuable camera had already all come off the end of the tracks onto the grass with a rather large jolt.

While I had been practising on the dolly, the boys had been rehearsing. Suggesting this had been a bad move on my part, because after they had run up and down the bank three times – once fast, once slow and once somewhere in between – and we had finally got their speed to match that of the camera, they decided they were bored. When they realised that they were going to have to do it many more times, that they couldn't race, and that they would have to stay in the same positions (Will could not go in front of Joss), then there were tears. Will decided the pressure of stardom was all too much and he wanted to retire from acting there and then. There and then while the sun was at a perfect angle. He stood there in his wellies, with his baggy shorts and orange T-shirt covered in mud and grime, holding his net, the setting sun glinting on his jam jar, and cried. We wanted to hug him and we wanted to kill him.

Andy looked as if he might faint. We all tried wheedling and coaxing but it was no good. This actor was out of the movie. Charlie and I began to discuss in hushed tones if it was at all possible to achieve the same effect for the opening with one boy running along the riverbank, or whether the excitement and chatter of two was absolutely essential.

Meanwhile, on the opposite bank negotiations had started between agent (dad) and actor (son). I overheard the word 'sweets' then the reply 'Chewitts'. I'm sure I also heard the number fifty along with the word 'packets', and one glance at Andy's face convinced me that he was desperate enough. He knew the pressure we were under. Charlie, as a fellow dad, was wincing in sympathy. Desperation was not great if you wanted to stay in control. But then, as if the sun had come out from behind a cloud, the raindrops on Will's face disappeared and, as if by magic, he was once more entranced by the idea of running along the bank as many times as we wanted him to. A deal had been struck. Andy went back to his hiding place behind the tree, his shoulders drooping. As the father of four boys, he would be the first to admit it, he was a beaten man.

Thanks to the satisfactorily concluded negotiations we were soon racing through the shots. The light was beautiful, and after a while the boys began to enjoy themselves. To begin with, although Will was prepared to fulfil his side of the deal, it was clear that Andy had forgotten to insert a subclause about smiling, and that had to be encouraged more than a little. But when it came to the fishing part of the sequence Will had a great time, and we enjoyed filming both of them. They were so at home in the river they didn't even notice the kingfisher that came to join them in one really special shot.

With the sun too low to give us any more decent light we packed up. Back at the house, two filthy boys were dunked in a bubble bath, filled up with chicken and noodles, and were fast asleep five minutes into their journey home. Fingers crossed, we had our opening sequence, as we had always dreamed of it, in the can.

The next day, we were due to give the camera back. Charlie spent his final hours with it filming the finishing touches in overcast weather: wide shots of the sewage farm, some more underwater shots of trout. It was impossible to believe we would no longer be filming the events and characters we saw every day on the river. It was a concept we couldn't grasp.

It had been the worst summer for fourteen years, and on 1 September, in contrast to the previous year's late autumn, a chill was already in the air. There is a freshness to the morning that never fails to remind you, even if you left school years ago, that it is the start of a new school year. It is a slight snap, delicious as you take it in, that says exciting things are about to happen, that makes those nerve ends jangle, that invigorates the skin. It hints at Bonfire Night, Halloween, casseroles, jumpers and cosy nights in front of the fire, and whispers of the full-on winter splendour of Christmas, which for me comes with the extra treat of a long period of shopping.

This is part of the joy, for me, of living in Britain. The changing seasons. I love this cold reminder on my skin, the first time that a plume of mist comes out with my breath in the mornings when I open the door to call the cat in. How refreshing after the long, hot, dusty summer to have a change.

The sediment settles

At least that's the theory. But this year we had been conned. This year I was not happy to feel the cold twang on my skin. I was not ready. Our long, hot summer had been only two weeks and I had been praying for an Indian summer.

From the filming point of view it was a relief that for once the weather didn't matter. We had, officially, finished. Hot on the heels of the filming came the edit. This was the really scary bit for me, and the part of the process of which Charlie and I had least experience, so we were both nervous. Luckily, writing scripts and viewing and logging all the footage took our minds off it a little, but for the two weeks leading up to it I had a ball of lead growing in the pit of my stomach. It was another reminder of the start of a new term.

The camera had gone back to the BBC stores amid much talk from Charlie of relief. He mentioned freedom, the joy of living without a film camera, being able to lie in without feeling guilty, being able to relax and enjoy the view instead of worrying that he should be filming it.

And indeed, he was trying to relax. We had kept Will and Joss busy while cloud covered the sky on the day of their shoot. They had picked lots and lots of sloes for us, beautiful berries growing in the hedge around the paddock, black with a blue blush, and I had been looking forward to making lots and lots of sloe vodka. Just put sugar, sloes and vodka into a big bottle for six months, or longer if you can bear it, shake a little every so often, and then drink. You do need to make it within a few days of picking the sloes, though, before they go all wrinkly, and the only spanner in the works is having to prick each sloe several times with a pin before plopping it into the vodka and sugar mixture. This is time-consuming as each sloe is about the size of a grape and for each bottle and a half of vodka you need a kilo. That's a lot of sloes. I had planned to do it in front of *EastEnders* so that I wouldn't notice how long it took.

But Charlie spotted the sloes and offered his services, grinning and reminding me that he could do other things at home

235

now that he didn't need to be out filming all the time. It would be relaxing to undertake such a beautifully mindless task, he said. Within five minutes he had given up doing each sloe individually and invented a system with a nutmeg grater that pricked at least five sloes at once. Within another couple of minutes he had realised that if you held the next batch of five sloes in the right hand while rolling the present batch under the grater with the left, transfer time between batches could also be minimised. He worked like a demon, rolling and passing, rolling and passing, frowning, and focused on getting the job done as quickly as possible.

Fred watched intently from the other end of the table shovelling chicken nuggets and baked beans into his mouth and I laughed my head off. It would take this man a while to adapt to a relaxing lifestyle.

Within two days we had the camera back. There were a few more shots he had thought of. I resigned myself to the fact that this was going to be more like a weaning process than a clean break.

Chapter Nineteen
GOING WITH THE FLOW

Letting go

I now realise that when we stopped filming we started to
let go. It was like taking a child to hospital and allowing
the professionals to look after him, all the while slightly
bewildered that you do not have the skills to perform the
operation yourself.

Now other people were part of our small team.

Now the hours were regular and we had to go to town
to work. There was no Jamie turning up at the house at
all hours of the day and night, and I didn't trip over cables
every time I went outside my front door. The spare room
was no longer full of kit and tools and batteries on charge;
it could be used for guests again. The bacon started to
accumulate in the fridge, the demand for sustaining bacon
sandwiches having dropped off. Ours was beginning to feel
dreadfully like a normal house.

Our editor was the first new member of the team. His name was Nigel Buck, and he was based just down the road from the BBC in Bristol in a tall, Victorian building. Nigel is remarkably thin for a man who sits down all day, every day, watching monitors. He is quiet and focused and concentrates very hard on what you tell him.

On the first day of the edit, Charlie and I were bags of nerves. I couldn't shake the first-day-at-school feeling. There was so much to learn. I had no idea whether our plans to put the thing together would work and I really wanted Nigel to like us and to enjoy working on the programme. Above all, I wanted him to like the programme. As we crawled through the rush hour, Charlie and I chattered incessantly. We talked nonsense, laughing and joking and taking the mickey out of the 'hit parade' on the radio. It was all to hide our nerves. The car was full of tapes, from the small DV tapes with night after otter night carefully recorded, to larger Beta tapes for editing. There were piles of CDs with tracks marked which we were hoping to use. There were files crammed full of notes and our new Apple computer with all the tapes logged on it. We stopped for fresh croissants to take in for breakfast, it would be a treat for the first day. We giggled as we resisted the temptation to eat them all in the car before we got there. We were like overexcited children.

It felt somehow as though Nigel were going to see us naked. As if we would march in all smiles, share a nice breakfast and then strip off and say, 'What do you think? You've seen a few of these, is what we've got good enough?'

He would be the first stranger to take a good hard look at everything we had been doing for the last eighteen months, and his opinion mattered, not least because he did this kind of work all the time. It is in an editor's hands to make a good programme bad or a bad programme good. We had faith in Nigel but we didn't want him to have to do the latter. We didn't want him spending the next six weeks of his life going home to Molly, his other half, wiping his brow, groaning and saying, 'God knows how I am going to make a decent programme out of that lot.' Charlie was in denial – he pretended he didn't care what anyone thought. But after two years of making babies and programmes with him I was beginning to know better. He would be more floored than I would if this programme wasn't a success.

We burst into Nigel's offices hiding our nerves with croissants and tapes and promises of a nice lunch and CDs and joviality. He fed us coffee, which strung us up even more, and sat down to view what we had.

So that we could at least begin with something well planned we decided to start on the opening, for which we had a storyboard. It was straightforward and we hoped it would give us a feel for the rest of the programme. We spent the first few days working on it – choosing effects to make it look nostalgic, a typeface for the title, the music, how the pictures would merge into one another – and it worked. It worked so well that each time I listened to the music and watched the monitor my eyes filled with tears. It was the realisation of a dream, a dream that for so long had been so clear in our heads, and now it had happened it seemed just as good. The tears were of relief and love, but I always sat at the back of the room behind Charlie and Nigel so that they never knew. If I thought they might look round to ask me something I just looked down at my laptop.

It then took nearly a week for Nigel to familiarise himself with everything else we had shot. At the end of the week there was nothing, no conclusion to our fears, neither positive nor negative. Nigel does not react much. We thought he liked us, we hoped he liked the film, but he didn't rave. He offered his opinions on how it should be put together, but he didn't rave. He spent loads of his own time creating new sequences and rectifying some of our mistakes and tracking down music that we would never have thought of, but he never once raved. He ignored our rows and carried on working while we argued over him, never joining in or allowing himself to be drawn into taking sides. He was a sobering influence on us, but I never really knew — I never had the guts to ask — whether he was in fact going home in the evenings and saying to Molly, 'They really are very nice, but they can't make a film for toffee.'

There were others, too. Mike Gunton was spending more and more time with us, viewing what we were doing and keeping a close eye on the novices. His experience was a comfort. We had no idea if we were doing it right, and the further into the process we went the less sure we were that this film was any good. He always allowed us room for our ideas and never once do I remember him telling us that they were crap, even though we told him some of his were. Also, we had too many ideas. Four weeks into the six-week edit our fifty-minute programme was still two and a half hours long. After the viewing we sat in silence, in the small hot room with the blinds pulled down, and Mike was the only one not panicking.

Day after day, Nigel whittled and whittled while I whined and pulled my hair out trying to unravel the structure I had created so that some sequences could be cut. In desperation, we appealed to Mike to show it as a two-parter, but he just laughed, assuming we were joking. So we laughed too, pretending we had been. We lost the duck rape sequence but kept the ducklings falling off the weir; we lost the ducks on ice but kept the car under the bridge; we lost a trip all the way down the river but kept the wellies through the sparkling water.

We took it apart and put it back together so many times that, finally, one afternoon in the last week, Mike and Nigel had to send us home early so that they could make some editorial decisions without our emotional input. Like those parents with the child in hospital we wanted to sit outside the door and listen, but we did as we were told and left. All night we fretted that some of our favourite bits had been axed, but the next morning it turned out that they had been as flummoxed as we were and had only managed to trim a couple more minutes. But we did it. Methodically, Nigel tightened it up, day by day, minute by minute. In that six-week period the programme changed significantly, but when we were finished it was still close to our original vision.

When at last Nigel had made the final cut, said goodbye and moved on to another project, we moved to the sound department. Now we were in completely unknown territory, but Martyn and Neill had done their research, and we could only marvel at the sounds they amassed and the way they laid them down onto our film to reproduce the sounds of the river. We just sat and nodded and smiled as they played with their noises. I would never have guessed that such artistry was involved with sound if I hadn't seen the process for myself.

I came into my own when it was time to do the commentary. This was my area of expertise, this was what I knew best. Intonation and delivery I could do, but it wasn't me doing the job and Charlie had bottled out. Although the film was very much still the story of his relationship with the river, he had decided that he hated the sound of his own voice. We had tried and tried through the edit to convince him, but each time we attempted to get him to lay down a guide, he was so nervous that his voice went all high. In the end we all agreed that a falsetto Charlie was not the effect we were after and that maybe Charlie was right. He was not the man to do it after all.

We were reluctant to use an actor, who might sound too polished, and after much deliberation and auditioning we plumped for the reassuring tones of a doctor with a personal interest – Charlie's dad, John. I loved every minute of our day doing the voiceover. John was nervous to begin with, but Martyn and I gave him plenty of encouragement and he soon got the hang of it and began to enjoy it. Sometimes the script worked well out loud, other times we had to cut things – which I found painful. Mike made sure he was there to convince me if I started to dig in my heels too much. We were having such a nice time that we were a little put out when the alarms sounded and we had to down tools and evacuate to the local church while the bomb squad investigated a suspect package. The church laid on soft drinks and sand-wiches and most people seemed disappointed when we got the all-clear and had to go back to work. I couldn't wait. In spite of the bomb scare, in just one day the whole voiceover was done. Apart from a few bits and bobs the programme was complete.

It was the details that seemed to take up so much time – the paperwork, signing off the budget, clearing the rights to use the music – but finally there was nothing left on the list. Anna Kington, our production coordinator, moved on to other projects, as had Neill and Martyn and Mike and Nigel. Even Jamie was away, filming in Africa again. We had little to do but wait for transmission day, only a few weeks away.

All too soon we were gathered in our darkened living room for the end of the story. Through a haze I heard a BBC announcer say something about a film about a river. I had a lump in my throat as the familiar music began. The tears pricked my eyes, as usual, during the opening sequence and my heart was thumping. I was relieved the room was in darkness and no one could see my face. Charlie's hand squeezed my shoulder.

All around us were friends, family, neighbours and the people who had helped us make the film. They had all made the journey to this special house to sit themselves on cushions and chairs and each other for fifty minutes to find out just what it was we had been doing on our river for the last eighteen months. The house was full to bursting and it was wonderful. Mike had turned up an hour or so early without his usual wisecracks and looking green, suffering badly from some kind of bug. 'I couldn't miss it,' he said. 'This programme is too special.' We had sent him to bed for a while with a Lemsip and cracked open the champagne for everyone else. Now we called him downstairs to join us. His faith in the programme had at times been the only thing that had kept us going, and the fact that it was special to him, too, was more than we could ever have asked for.

The familiar scenes washed over me, the programme unfolded, and I found that for the first time ever I was not critical. I just watched it. I even began to enjoy it. Charlie and I were arm in arm in the darkness, watching our film. A part of our lives was over.

I suddenly realised something. It was now late November. Three years before, almost to the day, the man of my dreams and I had set out for Skye to film otters.

Since then I had rowed with this man, laughed with him, insulted him. I had learned from him and taught him. He had been a source of irritation, strength and inspiration. He had become my lover, colleague and friend. In those three years I had shared everything with him and our lives had changed in every way. I wondered what he was feeling and thinking as he watched his shots, our pleasures and pains, and listened to his father saying my words. Was he remembering freezing otter nights? Was he reminding himself to take the mickey out of Jamie for something? Was he thinking of the time we had bickered about the inclusion of that last shot? Had he worked out that this was an important anniversary? I wondered how many more years it would be before I knew what he was thinking. His hand squeezed my shoulder again and I looked at him. He was smiling at me. I still found it hard to stop gazing into his eyes.

All was quiet. Everyone was focused on the flickering screen in the corner. But we had created something else in the last three years, something that had softened my heart and tested every part of me and that had no intention of keeping quiet. True to his genes, Fred proudly began his own commentary. 'There's the river. There's Fred's ducks. Oh, an otter. That's Fred's bridge ...' It became more and more difficult to hear his grandpa as Fred added a line of explanation to each shot. Fred had clearly decided that our film only lacked one thing – him.

When we could hear it, Charlie and I exchanged glances over bits of commentary we had changed or shots we had argued over as John's mellifluous voice led us through the story of the river and its characters. I have never watched a programme with which I have been involved with such emotion. The lump stuck in my throat, the tears kept welling. I felt like such an emotional fool. Yet this had been so much more than a job. Passionate as I am about all the programmes I present, this meant so much more. Everything was in it: our hearts and souls, our days and nights, and the things we loved the most. And there it was on BBC2.

The fifty minutes were cruel in their speed. The credits began to roll up the screen, the kingfisher flew away down the river in the last shot, people cheered as our names appeared. The lights came on and removed me from the safety of darkness. The house felt suddenly full and noisy again – champagne was popped open, the phone began to ring and my time for reflection was over. And in other people's houses ... What?

Most probably went to bath the kids and put them to bed, or put the dinner on. I wondered if they would talk about the programme and what they would say. Had it transported them, just for a while, out of their everyday lives and showed them that there is a different world? One which moves to different rhythms. I wondered if, for those fifty minutes, we had made them care. I desperately hoped we had. The thing is, with television programmes, you never really know.

In a state of numbness, I watched my little blond man as he mingled happily at thigh height with all the people in his house, pulling on their clothes and asking them if they had seen the ducks and the otter on his river. I looked out through the open curtains into the inky blackness. The moon was bright on the water.

This wasn't just Charlie's river any more.